IDAHO

WILDLIFE VIEWING GUIDE

Leslie Benjamin Carpenter

FALCON PRESS

MAJOR PUBLIC LAND AND WILDLIFE MANAGEMENT AGENCIES IN IDAHO

 BUREAU OF LAND MANAGEMENT administers 11.9 million acres of public lands in Idaho, nearly 22.5 percent of the state. The BLM is responsible for managing public lands and its resources, under a multiple-use mandate, to best meet the needs of the American public. The BLM has recently initiated a "Watchable Wildlife" program which is designed to increase opportunities for viewing wildlife on public lands.

 The **UNITED STATES FOREST SERVICE** manages 21.8 million acres in Idaho for multiple use. These public lands provide wildlife habitat ranging from sagebrush flats to riparian bogs to alpine peaks. They support endangered woodland caribou and bald eagles, world class salmon and trout sport fisheries, abundant big game herds of Rocky Mountain elk, mule and white-tailed deer, and a variety of upland game birds. Opportunities for viewing are as diverse as the five congressionally designated wildernesses, two national recreation areas, and one national grassland within the 12 national forests in Idaho.

 The **U.S. FISH AND WILDLIFE SERVICE** administers 81,602 acres in Idaho that include six wildlife refuges, one waterfowl production area, and three fish hatcheries. The USFWS administers programs for threatened and endangered species, migratory waterfowl and inland fisheries management. Its mission is to conserve, protect, and enhance fish and wildlife and their habitats for the continuing benefit of the American people.

 The **U.S. BUREAU OF RECLAMATION** has responsibility for over 140,000 acres of water and 70,000 acres of Idaho land in a variety of natural environments. The lands and water provide significant and critical habitat for several species of raptors, including threatened and endangered species, big game, small game, upland birds, waterfowl, and nongame animals. These resources offer a vast spectrum of opportunity to observe and study wildlife in its natural habitat.

 The **NATIONAL PARK SERVICE** manages 74,368 acres in Idaho including two national monuments, one national historical park, and one national reserve. The NPS was established to conserve and protect resources unimpaired for future generations while providing for public use and enjoyment.

 The **IDAHO DEPARTMENT OF FISH AND GAME** manages over 200,000 acres including 28 wildlife management areas and fishing lakes and 25 fish hatcheries. The IDFG is responsible for preserving, protecting, perpetuating, and managing all Idaho wildlife. The department administers programs for fish and wildlife management and research, game regulations, law enforcement for the conservation of wildlife and game and nongame educational programs.

 The **IDAHO DEPARTMENT OF PARKS AND RECREATION** manages 46,515 acres, including 22 state parks. The IDPR is charged with the responsibility for the aquisition, planning, protection, operation, maintenance, development, interpretation, and wise use of areas of scenic beauty, recreation utility, historic, and archaeological or scientific interests.

Copyright © 1990 by Falcon Press Publishing Co., Inc.
Helena and Billings, Montana

Published in cooperation with Defenders of Wildlife

Design, typesetting, and other prepress work by Falcon Press, Helena, Montana
Printed in Singapore

Library of Congress Number 90-81921
ISBN 1-56044-021-X

Front cover photo: Bighorn sheep
   by Tim Christie
Habitat artwork on pages 10-15
   by Rene´ Eisenbart

CONTENTS

Significant financial and technical contributions to the research and development of this guide were made by the following:

Primary funding and/or contributor—
Hadley Roberts
Defenders of Wildlife
U.S. Forest Service
U.S. Fish and Wildlife Service
U.S. Bureau of Reclamation
U.S. Bureau of Land Management
Idaho Department of Fish and Game
Idaho Department of Parks and Recreation
Idaho Audubon Council
The Albertson Foundation

Secondary contributors —
The Gannett Foundation
The Peregrine Fund, Inc.
Idaho Transportation Department
Idaho Chapter of The Wildlife Society
Craters of the Moon Natural
      History Association

GOVERNOR'S LETTER

From the snow-crowned peaks of the Sawtooth Mountains to the deep walls of Hells Canyon, Idaho abounds with beautiful places for people to visit and excellent places for wildlife to live. Thriving wildlife makes the Gem State's scenic beauty come alive.

Long recognized for its rich array of game animals, this guide recognizes Idaho's vast potential for nonconsumptive wildlife recreation. Wildlife watchers can expect to see mountain goats high atop wilderness peaks; salmon powering up cold, clear rivers; and bald eagles soaring lazily over city parkland.

This complete guide to Idaho wildlife-watching is your invitation to discover nearly 100 places known for their abundant, unusual, or diverse wildlife resources. The sites were carefully selected by seasoned Idahoan animal watchers and wildlife biologists.

Enjoy exploring Idaho's scenic and living wild wonders.

Sincerely,

Cecil D. Andrus
Governor

Idaho Wildlife Viewing Guide
Project Committee members:

USFS—Kathy Lucich, Vickie Saab, and Tim Schommer; USFWS—Jay Gore; BLM—Allan Thomas and Mark Hilliard; USBR—Jim Budolfson; IDFG—Wayne Melquist; IDPR—Jack Lavin and Larry Mink; Idaho Audubon Council—Al Larson

INTRODUCTION

Idaho is a state rich in numbers and diversity of wildlife with 357 bird, 109 mammal, 22 reptile, 15 amphibian, and 68 fish species. The wildlife occupy vast expanses of uncrowded, unspoiled natural areas, because over two-thirds of the state is public land and Idaho's population is low.

This guide will direct you to the best wildlife viewing opportunities in Idaho. It is designed for both the casual and experienced wildlife viewer, for planning specific viewing trips, or simply carrying in your car to assist in finding wildlife as you travel. Idaho has some outstanding scenic wonders that also provide great wildlife viewing opportunities. For instance:

- Northern Idaho has a greater concentration of lakes than any other western state. Lake Pend Oreille, the largest of these lakes, supports up to 400 wintering bald eagles, over 100 pair of nesting osprey, and thousands of waterbirds.
- The Salmon River, home to bighorn sheep, mountain goats, and hundreds of other wildlife species, is the longest undammed river in the United States. It is bordered by the Frank Church River of No Return Wilderness, the largest wilderness in the lower 48 states.
- The Seven Devils mountain range, inhabited by easy-to-view mountain goats, towers high enough to overlook four states. Neighboring this range is Hells Canyon, the deepest canyon in North America.
- The Snake River Canyon, within the Snake River Birds of Prey Area is supports the densest concentration of nesting raptors in North America. More than 700 pairs represented by 14 species nest along this 80-mile stretch of river.
- Bruneau Dunes State Park contains North America's tallest single structure sand dune and it's a great place to view shorebirds, ducks, geese, and swans during the migration seasons.
- In Hagerman Valley the disappearing Lost River of east central Idaho re-emerges as canyon wall cascades and valley floor springs, providing open water habitat to thousands of wintering ducks, geese, and other waterbirds.
- At the City of Rocks National Reserve ancient rock monoliths tower over 60 stories high like a lost desert city. Combined with an uncommon Idaho habitat

The peregrine falcon is one of Idaho's most popular but rarely seen birds. It can perform aerial maneuvers with jet speed to catch even the swiftest of birds. TOM ULRICH

type of pinyon pine and Utah juniper, the "city" attracts many songbird species that are only seen in this part of Idaho.
- At Craters of the Moon National Monument deer, marmots, pika, grouse, golden eagles, and several reptile species inhabit dramatic basaltic formations of cinder cones and barren lava flows.

Because so many people combine wildlife viewing with other outdoor recreation, a special effort has been made to include sites accessible by a variety of means—from a car, on foot, by small boat (canoe or raft), large boat, from a bicycle, cross-country skis, and even horseback. Recreation symbols accompany each site description as a quick reference in identifying a specific type of access.

The following pages have helpful tips for using the guide and finding and viewing wildlife. We have included three wildlife community diagrams and some basic information on the importance of habitats to wildlife. Learning to recognize different habitat types and to appreciate the value of these types to wildlife will enhance your viewing experience.

Idaho offers several viewing sites with exceptional learning opportunities. The Sawtooth Fish Hatchery is a great place to learn about the life history of chinook salmon. The Boise Nature Center includes a river observatory with underwater viewing stations that reveal the life cycle of fish. Surrounding the observatory are naturalistic backyards, a butterfly garden, and plenty of information on how to attract wildlife to your backyard. Kathryn Albertson Park, also in Boise, was designed for wildlife viewing, especially birds. All three sites offer guided tours to school groups and individuals.

RESPONSIBLE VIEWING

Most birders, wildlife photographers, sportsmen and others who venture outdoors to see wildlife share a genuine concern for and appreciation of wildlife. Unintentionally, wildlife watchers can harm wildlife through direct disturbance, feeding the animals, littering, or disobeying laws. People may place themselves in danger by approaching wildlife too closely or may spoil the viewing experience of others. The following guidelines will help avoid harm to people and wildlife.
- Minimizing Disturbances to Wildlife—An important aspect of responsible wildlife viewing is allowing animals to carry out their normal behavior without interruption. Animals are very sensitive to human presence and will flee if approached too closely. This can lead to a number of problems including: using up valuable energy at a time when they may already be stressed by winter cold or limited food supplies, abandoning eggs or young that need to be kept warm or protected from predators, or injuring themselves in the process of fleeing.
- Watch for subtle signs of distress such as head raised, ears pointed in direction of the observers, skittish movements, or alarm calls.
- Follow cautions in the site descriptions for particularly sensitive species.
- Use quiet, slow movements to avoid scaring wildlife. A car or boat makes a great blind in which to hide yourself and you may actually see more by remaining in it.
- Keep far enough away from nests and dens to avoid disturbing breeding wildlife which are especially sensitive.
- Never chase, repeatedly flush, or attempt to capture animals. Harassing animals is punishable by state and federal law.
- Do not pick up apparently sick or orphaned animals. You should phone the local IDFG conservation officer to evaluate a distressed animal properly.
- Obey Posted Rules—Many of the sites in this book have rules posted that explain when and where people can go. Some of these rules are designed to avoid disturbing

wildlife during important periods such as the breeding season.

• Observe road closed signs, which are often posted on national forests roads. Closures give animals space to move about without being disturbed or threatened by vehicle traffic.

• Always obtain permission from landowners before entering private property.

• Never feed wildlife. Animals accustomed to being fed by or being close to humans have a good chance of being hit by a car, becoming a nuisance and having to be removed from the area, ingesting plastic and other litter that can seriously harm their digestive systems, or starving to death when that food source is no longer present.

• Keep pets leashed, do not allow them to chase or harass wildlife.

• Never litter or deface property or the natural environment.

• Avoid Dangerous Wildlife—Always keep a good distance between yourself and rattlesnakes, bears with cubs, rutting elk and moose (in the fall), and moose with calves. These animals may charge if threatened and have the ability to inflict serious injury. If entering grizzly bear country, talk to the local managing agency and familiarize yourself with bear safety.

• Also, you should avoid getting too close to tame wildlife such as squirrels, raccoons, or deer as they may become very aggressive if not fed.

• Respect the Rights of Others—Be considerate when approaching wildlife already under observation by other viewers or photographers. Moving too quickly or approaching too loudly may ruin the experience for others.

TIPS FOR VIEWING WILDLIFE

A successful wildlife viewing trip involves the right equipment, knowledge of how to find wildlife, and an understanding of viewing etiquette.

The most important equipment for wildlife viewing is a pair of binoculars. Higher powered spotting telescopes are useful in areas where the wildlife is expected to be a good distance away, but may be too heavy to carry on long hikes. Field guides can greatly enhance the experience, especially if you are learning wildlife identification.

Some sites in this book offer more reliable viewing opportunities than others. This is often a function of the wildlife population densities as well as animal habitats and behavior. For instance, waterfowl can be seen more predictably than owls or snakes, just as elk or deer are more dependable than bighorn sheep or mountain goats. Each of the hundreds of wildlife species in Idaho have different daily and seasonal activity periods. Animals in desert habitats may be inactive and out of sight during the warm portion of the day, or only active at night. In addition, many desert animals leave the area or burrow underground during the hot, dry summers.

Typically, the best time of the day to view wildlife is when they are feeding, mainly in the early morning and late afternoon hours. The best seasons for seeing large numbers of wildlife, especially birds, is during the spring and fall migrations. Idaho is in the Pacific Flyway, a wide path used by birds that nest in more northern country and winter in Idaho or further south. Wetlands, including lakes, rivers, marshes, and mudflats are good places to look for wildlife. In winter, look for open water as animals, especially waterfowl and wading birds, congregate in these areas. Foothills at the base of mountain ranges can harbor hoofed mammals such as elk, deer, and pronghorn that move downslope to avoid poor foraging conditions in the deep snow. Information on the seasonal differences in species presence at sites are noted in the text.

Awareness of habitats is a valuable tool for wildlife watchers. Simply stated, an animal's habitat is where the creature lives. Habitat provides the four basics for survival—food, water, shelter, and space. For example, river otters live in river habitat. Rivers provide fish to eat and soil banks to den in. Knowing that otters live in rivers would be a valuable tool for a wildlife watcher hoping to see otters.

All of the fish, amphibians, reptiles, birds, and mammals living in one habitat type can be referred to as that habitat's wildlife community. For example, the river habitat may support a wildlife community of steelhead trout, Pacific tree frogs, garter snakes, osprey, and river otters. In addition, that habitat type consists of a community of different kinds of plants. On pages 10 through 15 are examples of three basic wildlife communities in Idaho—coniferous forest, sagebrush grassland, and riparian zone.

Within a habitat type, special features that provide for the basic needs of an animal may be required so that a particular species can occupy the area. Important features might include snags (dead, standing trees), logs, rocky cliff areas, a stream or water source, a burrow in the ground or a cave. Looking for these types of habitat features when you visit the viewing sites will help you to spot wildlife.

Some types of animals live only in very few kinds of habitats, so they are called specialists. For example, mountain goats are specialists because they only inhabit rocky alpine habitat. On the other hand, generalists thrive in a variety of habitats. Such an animal is the coyote whose yipping and howling can be heard in wetland, grassland, forest, and urban areas. Because there are so many wildlife species in Idaho it would be impossible to list all of those that occur at each site. By becoming familiar with Idaho's habitats and the different species that live in them, you can project what you will likely see on a trip to a particular site. By doing a little pre-trip planning, you can avoid the disappointment of not seeing much.

HOW TO USE THIS BOOK

The color strips on the outside of the pages are keyed to each of the seven travel regions in Idaho. The sites are numbered consecutively from 1 to 94, with all sites in a region placed together and numbered from north to south. Maps of each region with the site numbers are at the beginning of each section.

Symbols of animals and plants are limited to seven per site and indicate either the species most likely to be present or those unique to the area.

The site descriptions describe each habitat and then identify some of the species found at the site. Only unique or representative species are listed, not every species that may occur at the site.

Directions are given based on the Idaho Official Centennial Highway Map. Forest, county, and local roads are described from Bureau of Land Management or Forest Service maps. Only towns named on the Idaho Centennial Official Highway Map are listed as the closest town.

Facilities are listed only if they are on the site. In some cases, rest rooms or hotels may be nearby. The handicapped symbol is used at sites that accommodate the physically disadvantaged on trails or at interpretive facilities.

Ownership signifies the agency or entity that either owns or manages the site. The names of private owners or organizations are not always listed. Private sites have been included in the book only with the permission of the owners. Please respect their rights when visiting these sites.

The phone number listed after ownership is the number to use if you have questions concerning individual sites. This is usually the managing agency or individual. Unless otherwise noted all phone numbers are in Idaho, area code 208.

Recreational symbols are listed to indicate various opportunities at each site that are conducive to wildlife watching. The managing agency can provide information on other types of recreational activities at the sites.

MAP INFORMATION

Idaho is divided into seven travel regions shown on this map. Wildlife viewing sites are numbered consecutively in a general pattern from north to south then east. Each region forms a separate section in this book, and there is a map of numbered sites at the beginning of each section.

HIGHWAY SIGNS

As you travel across Idaho, look for these signs that identify wildlife viewing sites. Most signs show the binoculars logo or the words "Wildlife Viewing Area" with an arrow pointing toward the site.

TRAVEL INFORMATION

For additional travel information, contact Idaho Travel Promotion, Department of Commerce, Statehouse, Boise, ID 83720, or call 1-800-635-7820. For current road conditions call 208-336-6600.

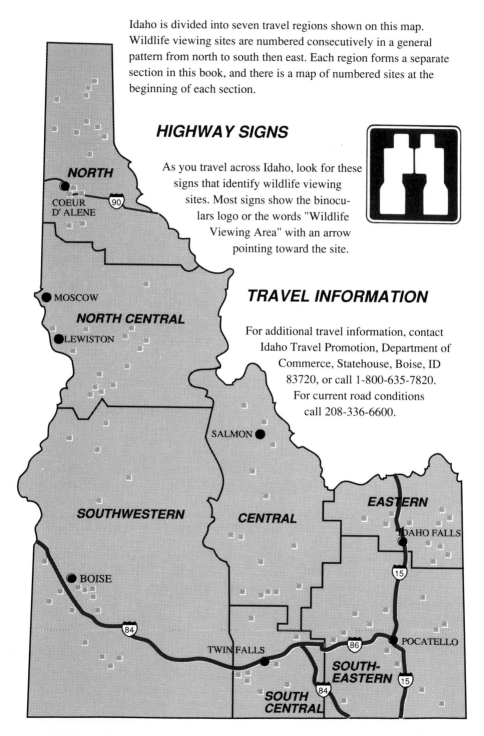

NORTH

COEUR D'ALENE

MOSCOW

NORTH CENTRAL

LEWISTON

SALMON

SOUTHWESTERN

CENTRAL

EASTERN

IDAHO FALLS

BOISE

POCATELLO

TWIN FALLS

SOUTH-EASTERN

SOUTH CENTRAL

FEATURED WILDLIFE

Carnivores

Songbirds

Insects

Freshwater mammals

Marine birds

Hoofed mammals

Fish

Shorebirds

Upland birds

Small mammals

Reptiles

Wading birds

Birds of prey

Wildflowers

Waterfowl

FACILITIES AND RECREATION

P Parking

Boat ramp

Campground

Handicap accessible

$ Entry fee

Small boat

Picnic

Hiking

Restrooms

Large boats

Bicycling

Cross-country skiing

Restaurant

Lodging

BEST VIEWING SEASONS

Spring   Summer   Fall   Winter

SITE OWNER /MANAGER ABBREVIATIONS

IDFG—Idaho Department of Fish and Game
USFWS—U.S. Fish and Wildlife Service
USFS—U.S. Forest Service
USBR—U.S. Bureau of Reclamation
BLM—Bureau of Land Management
ACE—U.S. Army Corps of Engineers

NPS—National Park Service
IDL—Idaho Department of Lands
ITD—Idaho Transportation Department
SP—Idaho Department of Parks
    and Recreation
PVT—Private

WESTERN TANAGER
Feeds on insects high up in the trees. Builds a nest of bark strips, grass, and twigs.

GOSHAWK
Chases small mammals and birds in thick forests. Short, rounded wings and a long rudder-like tail make it extremely quick and agile.

CONIFEROUS FOREST

Coniferous forests cover much of the mountainous northern half of Idaho. Interspersed with these forests are meadows, shrub fields, and riparian zones that provide habitat diversity essential to many wildlife species. Many animals, like elk and songbirds, must move to lower elevations or migrate out of the areas in the winter to find enough food.

DOUGLAS-FIR
Needles are an important winter food of blue grouse.

ELK
Feeds in open areas on grasses and the shoots, leaves, and buds of shrubs. Needs dense cover next to feeding areas for shelter, especially for calves.

BITTERCHERRY
Bright red fall berries provide nutrition for many birds and mammals. In addition, rodents and rabbits feed on its bark, twigs, and leaves.

MEADOW VOLE
Makes tunnels through grass between underground burrows and feeding areas. Is an important food for owls, foxes, coyotes, and other animals.

WESTERN HEMLOCK
Seeds are favored by chickadees and pine siskins.

PINE MARTEN
Preys on squirrels and smaller rodents in mature forest stands. Dens in hollow trees and rock caves.

RED SQUIRREL
Feeds largely on conifer seeds stored in forest floor caches. Nests in tree cavities and doesn't hibernate.

PILEATED WOODPECKER
Plays an essential role in nature by drilling cavities in trees that will be used by other birds and mammals for nesting. Feeds on insects, especially ants and beetle larvae.

LUPINE
Pollen provides bees with protein. It is an important food plant for many caterpillars.

BEAVER
Helps maintain a high water table in meadows by building dams. This encourages the growth of vegetation such as willows and sedges.

OSPREY
Hunts in dramatic fashion by plunging feet first to catch fish near the surface. Builds massive stick nests atop dead trees or other structures that allow a good view and easy access.

WILLOW SHRUB
Occupies areas with high water tables. Root structure stabilizes the stream bank, while the crown shades the stream, cooling the water to benefit fish.

TIGER SALAMANDER
Dwells in rotten logs, animal burrows, or other underground moist places. Serves as prey for various birds and mammals.

YELLOW WARBLER
Migrates from South America each spring to nest in willow or alder thickets. Feeds largely on insects.

CUTTHROAT TROUT
Native species feeds on aquatic insects in the stream and on terrestrial insects that fall into the water from streamside shrubs.

ASPEN

Sometimes associated with riparian zones. Provides food (bark) for beaver as well as structure (limbs) for beaver dams.

SCREECH OWL

Nests in tree cavities, abandoned woodpecker holes, and bird boxes near water. Has a varied diet of rodents, birds, and insects.

RIPARIAN ZONE

Riparian areas are the lush, green vegetation zones along creeks and rivers, around seeps, springs, lakes, and reservoirs, and in bogs and wet meadows. They are among the most productive habitats despite their relatively minor proportional occurrence. Some 75 percent of all vertebrates depend on the unique and diverse habitat niches found in riparian areas for at least a portion of their life requirements.

MOOSE

Forages on willows, other browse species, and aquatic vegetation in beaver ponds or lakes.

WATER VOLE

Strictly an alpine rodent, it feeds on blueberries and moist herbs. Builds nests under the snow.

GOLDEN EAGLE
Raises its young in large stick nests built on cliffs, trees, and transmission towers. Rabbits and squirrels make up a large part of its diet.

BLUEBUNCH WHEATGRASS
Provides nesting and hiding cover for reptiles, small mammals, and birds. Many herbivores graze on it.

TOWNSEND'S GROUND SQUIRREL
A very important prey item for a number of mammals and a variety of raptors, especially prairie falcons. They stay underground from mid-July to January.

SAGEGROUSE
Uses sagebrush for food and cover. Gathers on traditional display grounds, called leks, for spring mating.

BADGER
Digs out meals of ground squirrels, gophers, and mice with its powerful front legs and long claws.

WINTERFAT
An important winter survival plant for pronghorn and deer.

SAGEBRUSH GRASSLAND

The sagebrush grasslands that cover nearly one-half of southern Idaho are part of the Great Basin desert. They are characterized by a simple habitat of shrubs and grasses, short, hot summers and very cold winters. Wildlife cope with the extreme warm conditions by migrating away from the area for part of the year. Non-migratory animals are often nocturnal or estivate underground during warm periods.

PRONGHORN
Forages on sagebrush, grasses, and desert plants. Able to run over 40 miles per hour, it is the swiftest North American mammal.

G SAGEBRUSH
vides food or cover
most species of
life living
his habitat.

HORNED LARK
Nests on the ground and feeds largely on insects. A very common bird of open habitats.

GOPHER SNAKE
Kills its prey of rodents and birds by constriction, then swallows them whole. Common in many habitats. Young hatch in communal nests.

SITE	SITE NAME
1	**Purcell Mountains Area**
2	**Kootenai National Wildlife Refuge**
3	**McArthur Wildlife Management Area**
4	**Priest Lake State Park**
5	**Albeni Falls Dam and Vicinity**
6	**Round Lake State Park**
7	**Sandpoint City Beach**
8	**Lake Pend Oreille—Pack River Delta**
9	**Lake Pend Oreille—Clark Fork River Delta/Fish Hatchery**
10	**Farragut State Park**
11	**Templin's Resort**
12	**Lake Coeur d'Alene—Tubbs Hill Nature Park/Cougar Bay**
13	**Lake Coeur d'Alene—Wolf Lodge Bay**
14	**Coeur d'Alene River—Chain of Lakes**
15	**Heyburn State Park**

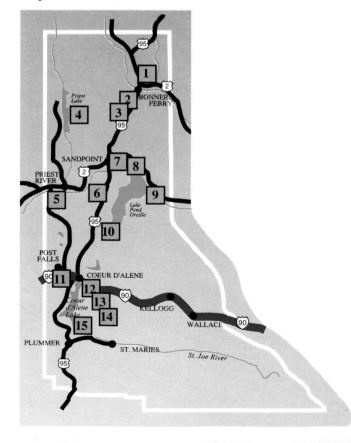

Description: This scenic route through the Purcell Mountains leads you to four mountain lakes, past a prairie marsh, along the Moyie River, and up to an alpine lake. The lower lakes offer good opportunities to see moose, wood ducks, common goldeneyes, mergansers, and osprey. While driving by Round Prairie, a marshy meadow along U.S. 95, look for waterfowl and wading birds. Take the short loop trails from the Robinson Lake day use area to view a heron rookery and osprey nest. Meadow Creek Campground has hiking and mountain bike trails along the Moyie River; look here for riparian songbirds and white-tailed deer. Queen Lake is a high mountain lake with moose, boreal chickadees, and boreal owls. Although owls are not likely to be seen, they will respond to imitated calls at night. Most of the lakes have camping and/or day use areas, fishing, and small boating opportunities.

Directions: This 40-mile loop tour begins and ends just north of Bonners Ferry, taking several detours off the main roads. From Bonners Ferry, go north four miles on U.S. 95 and turn right onto Forest Road 1005 to Smith Lake. Return to U.S. 95, continue 11 miles north, and take the Brush Lake turnoff, one mile past Idaho 1. Brush Lake is two miles in on Forest Road 1004. Return to U.S. 95, proceed east for eight miles, then take the Robinson Lake turnoff (you will pass Round Prairie in five miles). Return again to U.S. 95, continue east for two miles, and turn right onto gravelled Forest Road 211, just past Good Grief. Meadow Creek Campground is in 10 miles, just after the intersection with Forest Road 229. Return to Road 229, follow for 3.5 miles, and take the turnoff for Queen Lake (dirt Forest Road 2542). Drive seven miles in, then walk a quarter mile trail to reach the lake. Continuing on Forest Road 229, Dawson Lake is in two miles and the junction with U.S. 95 is five miles beyond that. Roads are not winter-maintained; the best viewing is from late spring through fall. Information and area maps are available at the Bonners Ferry District Ranger Station. If you have time you may want to visit the scenic Copper Falls, which features a loop trail that is accessible to the physically disadvantaged. The turnoff for the falls is three miles north of Good Grief.

Ownership: USFS (267-5561), PVT
Size: 40-mile loop **Closest Town:** Bonners Ferry

Standing dead trees, or snags, provide a portion of the life support system for many animals. They are used for perching, feeding on insects, and nesting and roosting sites.

2   Kootenai National Wildlife Refuge

Description: The wide variety of habitats in this scenic refuge support abundant and diverse wildlife. Meadows are interspersed with grain fields and wetlands in the valley bottom adjacent to the Kootenai River. Wetlands feature open-water ponds, cattail marshes, tree-lined ponds, and rushing creeks. A small portion of the refuge ascends the foothills of the densely forested Selkirk Mountains. Approximately 218 bird and 45 mammal species are found on the refuge. Tundra swans, Canada geese, and ducks are most abundant during spring and fall migrations. Common summertime birds include the great blue heron, Canada goose, osprey, northern harrier, ruffed grouse, great horned owl, and numerous songbirds. Bald eagles, which nest on the refuge, and rough-legged hawks are present in higher numbers from November through March. Look for elk, white-tailed deer, moose, beaver, coyote, and black bear during the fall and spring—especially in the morning and evening hours. View the refuge via a 4.5-mile auto tour, or from 5.5 miles of walking trails. There are mountain biking trails to the west of the refuge; check with the Forest Service for the best routes. Use of the area is restricted during the fall waterfowl hunting season on weekends, Tuesdays, and Thursdays. For additional viewing, head north to Canada and stop by Creston Wildlife Management Area, ten miles north of the Idaho border. It has a visitor center and great wildlife viewing opportunities.

Directions: *Take U.S. 95 to Bonners Ferry, then follow Riverside Road west for five miles to the refuge entrance. The refuge office (open Monday through Friday) is two miles beyond the entrance. Brochures, maps, and a wildlife checklist are available at the office and at several boxes on the refuge.*

Ownership: USFWS (267-3888)
Size: 2,774 acres   **Closest Town:** Bonners Ferry

Woodland caribou inhabit remote fir and spruce forests of the Selkirk Mountain Range. This federally endangered race of caribou is monitored by wildlife biologists concerned with its status. TOM ULRICH

3	**McArthur Wildlife Management Area**

Description: This WMA, with 600 acres of marshy lake surrounded by very scenic coniferous forest, was the state's first land acquisition for waterfowl production. More than 400 Canada geese were produced here in 1989. Over 150 nesting platforms and 30 artificial islands help increase duck and geese numbers. Visitors should stay away from the nesting platforms (raised wooden poles with straw bales) between the March 15 through June 15 nesting season. The best visiting period is late spring through summer, when coyotes, moose, and white-tailed deer are often seen at dawn and dusk. Hunting and fishing are popular here; boating, with a 10-horsepower limit, is allowed from July 1 to March 15.

Directions: From Sandpoint, follow U.S. 95 for 17 miles north (or from Naples, 3.5 miles south), then take County Road A4 west for 0.1 mile to the WMA parking lot. A brochure is available at the WMA and at the IDFG office in Coeur d'Alene.

Ownership: IDFG (765-3111)
Size: 1,700 acres **Closest Town:** Naples

4	**Priest Lake State Park**

Description: This large, pristine lake is surrounded by densely forested mountains of the Selkirk Range. From the northern edge of the park to Canada the Selkirks provide habitat for small numbers of the federally endangered grizzly bear and woodland caribou. Follow the rugged roads that originate from either of the two park units and head east into the mountains, where you may see white-tailed deer, moose, black bear, coyote, or, rarely, mountain goat. The uncommon Harlequin duck, which nests on undisturbed pristine streams, can be found in the upper Priest River drainage. The park features opportunities for rugged backcountry exploration as well. A 4-wheel drive vehicle is recommended on the roads. For the best viewing, follow the south and north fork trails of Indian Creek, Lion Creek, and Caribou Creek during spring, fall, or winter.

Directions: This park is on the eastern shore of Priest Lake. Take Idaho 57 north to Dickensheet Junction. Turn east and continue past the town of Coolin to the Indian Creek unit of the park, 10 miles north of Coolin. You can continue north another 11 miles to the Lionhead Unit of the park. Brochures and guided walks (call ahead from mid-September to late May) are available at both sites. USFS maps are available at the Priest Lake Ranger District on Idaho 57 on the lake's west side and at the Indian Creek park office.

Ownership: SP (443-2200)
Size: 418 acres **Closest Town:** Coolin

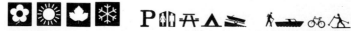

5   Albeni Falls Dam and Vicinity

Description: This dam retains 25 miles of the Pend Oreille River below the large and highly scenic Lake Pend Oreille. From the dam vista point look upstream to the railroad bridge that supports one or two osprey and Canada goose nests. In the winter, bald eagles are often spotted from this point. From the powerhouse parking lot at the top of the dam, scan the trees on the high bluffs for wintering bald eagles and look for osprey in the pilings downstream of the dam. Army Corps of Engineers recreational sites along the river edge offer additional viewing. The facilities are open from May through September, although they are accessible on foot year-round. The Priest River site is great for viewing osprey at their nests; you also may spot a beaver dam. Riley Creek is a good place to see white-tailed deer and, on occasion, moose. Albeni Cove shelters waterfowl in the spring and fall, where osprey can be seen in the summer and bald eagles in the winter. Although the road to the dam is closed in winter, you can walk in. Painted turtles, the only turtle species in Idaho, can often be seen basking on logs or rocks in ponds next to U.S. 2 between Sandpoint and Newport. From December to March, look for bald eagles perched in large cottonwood trees and snags along this same stretch of road.

Directions: From Priest River, follow U.S. 2 west for four miles and turn left at the sign for the dam. There are two viewing locations near the dam: the popular vista area sits on a bench of land above the dam and provides an upstream view, while the parking lot at the top of the dam provides both up and downstream views. Additional viewing can be found at the following ACE recreational areas: Priest River (on U.S. 2 just east of Priest River), Riley Creek (one mile south of Laclede off U.S. 2), Albeni Cove (continue west on U.S. 2 past the dam to Newport, then follow Idaho 41 for 1.5 miles until the road curves to the right (south), then turn off onto a county gravelled road and follow for 1.3 miles). Maps of the area are available at the vista area.

Ownership: ACE (437-3133)
Size: 13 miles of river **Closest Town:** Priest River

Beavers are the largest North American rodents. Though not often seen, a sharp-eyed observer may find a beaver-carved tree stump or beaver dam. JEFF FOOTT

6 | Round Lake State Park

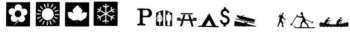

Description: This 58-acre, glacially created pothole lake is encircled by coniferous forest. The lake supports nesting osprey and great blue heron, beaver, muskrat, mink, and a few species of amphibians and reptiles. White-tailed deer, raccoon, and red squirrel are among the mammals you may see while exploring the park's trails. Numerous bird species include northern flicker, pileated woodpecker, barn and tree swallows, gray and Steller's jay, mountain bluebird, cedar waxwing, and western tanager. A foot trail around the lake and along the riparian tree-lined Cocolalla Creek takes you under canopies of western white pine, Engelmann spruce, grand fir, lodgepole pine, black cottonwood, paper birch, red alder, and Rocky Mountain maple. A self-guided botanical trail includes a view of the state flower, the syringa, which blooms in late June. Wildlife can be observed year-round although spring and fall are best. Ducks and geese may use the lake until it freezes in the winter, when cross country skiing is possible. For additional viewing, the Morton Slough area, where Cocolalla Creek joins Lake Pend Oreille, has shallow marshes that provide great habitat for waterfowl and songbirds. Osprey, bald eagles (in the winter), and golden eagles are often seen in this area.

Directions: From Westmond travel 1.5 miles north on U.S. 95. Turn west onto Dufort Road and drive two miles to the park entrance. If you are travelling from Sandpoint, follow U.S. 95 south for 8.5 miles to the turn off. There is a parking lot adjacent to the picnic area. The visitor center has wildlife, botanical, and geological displays, brochures, and trail maps. To reach Morton Slough, continue past the park on Dufort Road for another four miles. View here, at the lake edge, or head north for 1.5 miles where the road crosses over the slough. Follow this road another nine miles to return to U.S. 95 at the Osprey Nests Viewpoint.

Ownership: SP (263-3489)
Size: 200 acres **Closest Town**: Westmond

The brilliant yellow warbler is a familiar nesting bird in Idaho's riparian areas and shrub thickets. Because their insect prey becomes scarce in winter, these warblers migrate to Central and South America. TOM ULRICH

7 Sandpoint City Beach

Description: This peninsula of sandy beach and mud flats on the edge of Lake Pend Oreille is a "trap" for many migrating bird species. During spring and fall migrations many gulls, terns, and shorebirds use the area as a rest stop. For additional viewing, April and May visitors may want to stop by Grouse Creek Falls, an excellent opportunity to view rainbow trout jumping the falls on the way to their upstream spawning habitat. The USFS has constructed a quarter-mile trail to the falls.

Directions: From downtown Sandpoint follow signs to the beach. Cross Sand Creek and watch the shoreline along the beach, especially on the north side. To reach Grouse Creek Falls, follow U.S. 95 from Sandpoint for seven miles north to Colburn. Turn right onto Colburn-Culver Road and continue for four miles. Turn left onto Grouse Creek Road (Forest Road 280) and go five miles to the trailhead parking lot.

Ownership: City of Sandpoint (263-2123) [Grouse Creek: USFS (263-5111)]
Size: 200 acres **Closest Town:** Sandpoint

8 Lake Pend Oreille—Pack River Delta

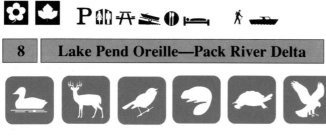

Description: Lake Pend Oreille, measuring 43 miles long and over 1,000 feet deep, is completely encircled by mountains and bordered by several wildlife management areas and undeveloped public land. The lake becomes a haven for wildlife during the fall and spring; wildlife is less abundant during the summer and winter. In the delta look for waterfowl (including Pacific, red-throated, and common loons), white-tailed deer, moose, beaver, and muskrat. Osprey, bald eagles, and great blue heron nest along the shoreline. On the west side of the highway bridge is a slough where moose are so frequently seen the state highway department has installed special highway signing.

Directions: There are two access points to the Pack River area. From Sandpoint, take Idaho 200 east for five miles. Turn east at the Sportsman's Access sign and follow the road for four miles to its end at Sunnyside. Wildlife can be viewed from any safe turnout along this road. You can also walk north from Sunnyside along a primitive county road paralleling the west shore of Pack River Delta. Boat access is best by canoe due to the shallow water and presence of stumps. Returning to Idaho 200, continue east for four miles to the bridge over Pack River Delta. Pack River Flats is on the north side of the bridge. Continue along the lakeshore on Idaho 200 for approximately three miles, stopping at the frequent view points. At the end of the route is Trestle Creek Recreation Area with picnic sites, a boat launch, and restrooms. A map of the area is available at the USFS office in Sandpoint.

Ownership: ACE, managed by IDFG (765-3111)
Size: Eight miles long **Closest Town**: Sandpoint

9   **Lake Pend Oreille—Clark Fork River Delta/Fish Hatchery**

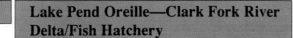

Description: This is one of northern Idaho's best areas for viewing waterfowl, osprey, bald eagles, and shorebirds. The Clark Fork River Delta is a mixture of cottonwood riparian forest, open water, grassland, small lakeshore wetlands, and exposed mudflats. High numbers of waterfowl and shorebird species are present during spring and fall migrations, as well as many songbirds. Look for redheads, wood ducks, and tundra swans. Over 20 pairs of osprey and several great blue heron nest in the vicinity, which is an important wintering area for up to 400 bald eagles. Elk, white-tailed deer, and moose are common. The Clark Fork hatchery offers guided tours seven days a week, featuring several fish species including west slope cutthroat, brook, brown, rainbow, and golden trout, and kamloops. Most fish species are on-site in the summer, while cutthroat and kamloops are present in the winter. In April wild kamloops, and from November to December wild kokanee, spawn in the creek next to the hatchery. The hatchery area also has a nesting osprey pair, wintering bald eagles, white-tailed deer, and moose. The uncommon harlequin duck can sometimes be seen in May on the creek.

Directions: See directions for Pack River Delta. From Trestle Creek Recreation Area, continue southeast on Idaho 200 for seven miles. You will see Denton Slough to your right. For the next five miles view from the road. To access the delta, drive to Clark Fork, turn right off Idaho 200, and cross the bridge over the Clark Fork River. Turn right and go two miles to Johnson Creek Recreation Site. There are no developed trails but you can walk along the dirt spurs into marshy areas. Canoeing is the best way to explore the delta but should only be done outside of the nesting season. To reach the state fish hatchery turn left off Idaho 200 north of Clark Fork onto Spring Creek Road, immediately before the Lightning Creek Bridge, and drive 1.8 miles. Food and lodging is available in Clark Fork. The campground at Johnson Creek is undeveloped.

Ownership: IDFG (765-3111), ACE (437-3133), PVT
Size: Seven miles of river   **Closest Town**: Clark Fork

The uncommon harlequin duck may be sighted on Spring Creek near the Clark Fork Hatchery, on the Upper Priest River drainage, and on the Lochsa and Selway rivers.
TOM ULRICH

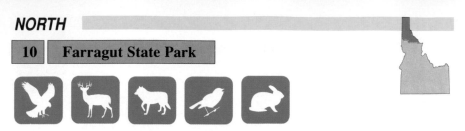

10   Farragut State Park

Description: This park offers scenic views of tall forested mountains, steep rugged cliffs, and Lake Pend Oreille. A unique wildlife attraction is the mountain goats that roam the steep cliffs of Bernard Peak, across the lake. The park has an outdoor mountain goat display and binoculars for public use at the Willow Picnic Area. The easiest way to view the goats is with binoculars from the lakeshore. For a closer look you can approach the cliffs by boat. The goats have become quite accustomed to boats and will not flee as long a you sit quietly. Park trails lead through forest habitat where white-tailed deer, red squirrels, Columbian ground squirrels, chipmunks, and songbirds are common. Black bear, mountain lion, bobcat, and coyote are also present, though rarely seen. Winter is a great time to view bald eagles.

Directions: From U.S. 95 at Athol, turn east onto Idaho 54 and go four miles to the park entrance. At the headquarters you'll find nature and historical displays, maps, and information on park activities.

Ownership: SP (683-2425), IDFG
Size: 4,000 acres **Closest Town:** Bayview

11   Templin's Resort

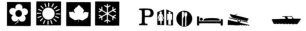

Description: Situated alongside the scenic Spokane River, this site offers a unique opportunity to see wildlife while enjoying the comfort of a vacation resort. View the river area from inside the restaurant. Osprey, great blue herons, several duck species, and tree and violet-green swallows are commonly seen in the summer. Look for bald eagles in the winter and California and ring-billed gulls, river otter, and mink year-round.

Directions: From Coeur d'Alene, take Interstate 90 west to Post Falls. Exit at Spokane Street and head south. The resort is located at 414 E. 1st Avenue.

Ownership: PVT (773-1611)
Size: 22 acres **Closest Town:** Post Falls

Mountain goats have two toes on each of their hoofs. The hoofs also have soft pads on the bottom that help in climbing around rocky ledges.

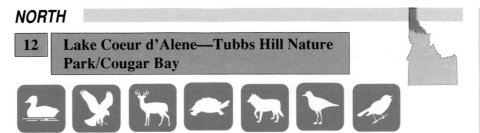

| 12 | Lake Coeur d'Alene—Tubbs Hill Nature Park/Cougar Bay |

Description: Tubbs Hill is a forested peninsula that extends nearly a mile and a half into picturesque Lake Coeur d'Alene and features numerous trails. The main trail begins at the public boat launch on the peninsula's west side and follows the shoreline for 1.4 miles to Tubbs Point, a rocky jut bordered by beaches, continues around the point, and passes three-story tall caves, more beaches, scenic overlooks, and side trails that go over the top of the hill. Exit at the marina or continue to McEuen Playfield, next to the boat launch. This site is great for watching numerous forest songbirds. In the summer look for black-capped, mountain, and chestnut-backed chickadees, red-breasted nuthatches, Steller's and gray jays, and evening and black-headed grosbeaks. In the winter watch for pine siskins, dark-eyed juncos, red crossbills, and downy woodpeckers. Commonly sighted lake birds are the osprey, common merganser, and pied-billed grebe in the summer and bald eagles, hooded mergansers, and horned, eared, and western grebes in the winter. Cougar Bay, two miles to the west, is a shallow bay rich with aquatic vegetation and bordered by marshland, coniferous forest, and fields that attract an abundance of wildlife. During winter and spring migration the bay is filled with waterfowl including Canada geese, blue-winged and cinnamon teal, shovellers, ruddy ducks, and mallards. In the summer breeding birds include osprey, great blue heron, red-necked and pied-billed grebes, cinnamon teal, sora rail, killdeer, spotted sandpiper, and common snipe. Western toads and Pacific tree frogs croak loudly at night. The proximity of forest and farmlands provides year-round viewing of songbirds including warblers, chickadees, flycatchers, and woodpeckers along with raptors, elk, deer, coyote, and black bear. Bald eagles are common in the winter.

Directions: Take Interstate 90 to Coeur d'Alene and exit onto Business 90 (Northwest Blvd.) south. Turn right onto Front Street and follow it to a city parking lot at Fifth Street. You will see McEuen Playfield and a public boat launch. The main trailhead begins at the launch and ends at McEuen Playfield, near the boat launch. A map at the trailhead shows all of the trails. Coeur d'Alene, within easy walking distance, has all facilities. Cougar Bay is two miles southwest of Coeur d'Alene on U.S. 95. You can view from the road for the first mile and then turn off onto Cougar Gulch Road and Meadowbrook Road. Park in the wide road turnouts. This area is privately owned, so stay on the roads and walk past the marsh and field areas.

Ownership: City of Coeur d'Alene (667-9533), PVT
Size: 150 acres, two miles of shoreline **Closest Town**: Coeur d'Alene

The grizzly bear is one of Idaho's rarest large mammals. Despite hunter education, a major threat to the species' survival is illegal shooting.

| 13 | Lake Coeur d'Alene—Wolf Lodge Bay |

Description: Wolf Lodge Bay attracts up to 60 migratory bald eagles when the lake's kokanee salmon begin to spawn and die during November. The surrounding steep mountain slopes, covered with dense stands of western larch, Douglas-fir, ponderosa pine, and grand fir, provide excellent communal roosting perches for the eagles. Good viewing areas are at Higgins Point, Mineral Ridge Boat Launch, and Mineral Ridge Trailhead. Peak populations are generally from late December to the first of January; by February most of the eagles have dispersed from the area. Bald eagles are sensitive to human disturbance, especially when approached on foot; stay in your boat or near your car. From September to November view migrating chinook salmon from the frontage road bridge over Wolf Lodge Creek just north of the Interstate 90 bridge crossing. Spring and summer visitors can take a three-mile loop nature trail at Mineral Ridge to see warblers, thrushes, vireos, and chickadees. The marshy areas at the mouths of Wolf Lodge and Blue Creek are also good birding spots—view from the road only. There is an additional hiking trail (FS 257) off Beauty Creek. To reach the trailhead go 2.5 miles northeast on Forest Road 438 (just past Mineral Ridge Campground). On the south side of Wolf Lodge Bay, adjacent to Idaho 97, Coeur d'Alene salamanders—endemic to northern Idaho and western Montana and an IDFG species of special concern—can be found on moist talus rocks at night.

Directions: From Coeur d'Alene, take Interstate 90 east for six miles to Higgins Point; look for bald eagles from this turnout. Continue around the bay by taking the bridge over Wolf Lodge Creek at the Idaho 97 junction. Use established turnouts to view wildlife. A detailed brochure on the eagles is available from the BLM office at 1808 N. Third Street in Coeur d'Alene.

Ownership: BLM (765-1511), PVT
Size: 900 acres **Closest Town**: Coeur d'Alene

Large numbers of Canada geese nest in Idaho and are easily seen at many of the viewing sites. TIM CHRISTIE

| 14 | Coeur d'Alene River—Chain of Lakes |

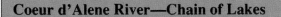

Description: This 13-mile stretch of the Coeur d'Alene River is bordered by numerous shallow lakes and marshy wetlands—an excellent area for canoeing. Adjacent mountain slopes are thickly covered with coniferous trees and dense shrubby undergrowth. Ducks, Canada geese, and tundra swans are numerous, particularly during migration. It is a major breeding area for tree cavity nesting wood ducks. Osprey, great blue heron, American kestrel, red-tailed hawk, belted kingfisher, and several species of swallows are also very visible. Muskrat and beaver lodges dot the marshlands and white-tailed deer and elk can be seen in the early morning and evening hours. Most facilities are available at Thompson, Cave, Killarney, and Rose Lakes and at Rainy Hill. Fishing and hunting are popular area activities. An additional viewing area is Old Mission State Park, which features the oldest standing building in Idaho. The park sits atop a high knoll overlooking the Coeur d'Alene River where waterfowl, great blue heron, osprey, white-tailed deer, elk, muskrat, and beaver are often seen from spring through fall. Bald eagles use the river area in the winter. There is a park exhibit on the great blue heron.

Directions: From Coeur d'Alene, take Interstate 90 for 18 miles east to the Rose Lake interchange, then turn south on Idaho 3 to Rose Lake. The Chain of Lakes is scattered along both sides of the highway for the next 13 miles. IDFG headquarters is at Thompson Lake at the end of the route. Wildlife viewing is from Idaho 3, its many side roads, or from a small boat. A brochure, map, and wildlife checklist are available at the Thompson Lake headquarters. To reach the Old Mission State Park, continue east on Interstate 90 for four miles east of the Rose Lake interchange and take Exit 39.

Ownership: IDFG (765-3111), USFS, BLM, PVT [Old Mission State Park (682-3814)]
Size: 13 miles of river **Closest Town:** Harrison

Wood ducks are strikingly colored residents of forested streams and ponds. They raise young in tree cavities and nest boxes. A good place to see them is the Chain of Lakes region. TOM ULRICH

15   Heyburn State Park

Description: This large, forested state park lies along the south shore of Lake Chatcolet. Here the St. Joe River meanders between two lakes toward Lake Coeur d'Alene. Extensive marshes and riparian-lined shallow lakes provide important stopover habitat for waterfowl and nesting habitat for ducks, great blue heron, and osprey. The area hosts over 50 pairs of osprey and a great blue heron rookery, which is near Benewah Lake. The park provides osprey, wood duck, and Canada geese nest structures, which has helped increase their numbers. The area is also well-known for its abundance and diversity of songbird species. A good way to see them is on one of the six forest trails (a total of 20 miles available), many of which are shaded by 400 year-old ponderosa pines. Mammals commonly seen include the red squirrel, yellow-pine chipmunk, snowshoe hare, raccoon, muskrat, beaver, badger, striped skunk, river otter, coyote, white-tailed deer, bobcat, and black bear. Present but rarely seen are the northern flying squirrel, marten, wolverine, mountain lion, and elk. Traveling the St. Joe River by canoe or kayak is also a great way to see area wildlife.

Directions: From U.S. 95 at Plummer turn east onto Idaho 5 and go five miles to the park entrance. A brochure and information on local attractions is available at the park headquarters (closed winter weekends). Campfire programs and guided hikes are conducted from Memorial Day through Labor Day. USFS maps are available in St. Maries, 13 miles east of the park.

Ownership: SP (686-1308)
Size: 7,838 acres **Closest Town:** Plummer

Though seldom seen, black bears are fairly common residents of many Idaho habitats. They are solitary, nocturnal, and naturally shy of humans. DENNIS AND MARIA HENRY

SITE	SITE NAME
16	Mallard-Larkins Pioneer Area
17	Hells Gate State Park
18	Craig Mountain Wildlife Management Area
19	Winchester State Park
20	Nez Perce National Historical Park— Heart of the Monster Unit
21	Musselshell Meadows
22	Lochsa River Canyon
23	Selway River Canyon
24	Elk City Area
25	Lower Salmon River Canyon
26	Middle Salmon River Canyon
27	Snake River in Hells Canyon National Recreation Area
28	Seven Devils Mountains in Hells Canyon National Recreation Area
29	Rapid River Fish Hatchery

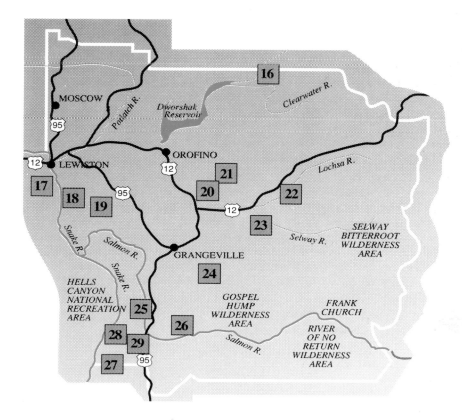

16	Mallard-Larkins Pioneer Area

Description: The Mallard-Larkins Pioneer Area includes about 30,000 acres of high elevation lakes and coniferous forest. It is part of a much larger area under consideration for wilderness classification which extends from the Little North Fork to the headwaters of the North Fork Clearwater River and encompasses 280,000 acres. There are about 280 miles of trails in the entire roadless area. There are two main entry points from the North Fork Clearwater River. Information on other entry points is available from the North Fork and Avery Ranger District offices. The most popular trail is Trail 420, along Smith Ridge, which begins at Forest Road 700. This is a high elevation trail that leads to the Larkins Peak area and to a group of subalpine lakes. It is about five miles in to the first lake. The trail is usually not snow free until early July. Mountain goats are often seen near Larkins Peak and Heart Lake. The second entry point is from Isabella Creek on Trail 95, which begins at Forest Road 705. Trail 95 will intersect with Trail 96 about 1.5 miles in. Follow Trail 96 for one mile to visit a stand of very old western red cedars and western hemlocks known as the Heritage Cedar Grove. Turn around and continue on Trail 95, which follows a comfortable grade to a ridge top junction with Trail 97 (about five miles in). From here you can go north to Mallard Peak and the lakes or south toward the Black Buttes Area, which can be part of a loop hike. Either direction has spectacular scenery. If you go south, take the trail to Black Mountain, about three miles from the junction of Trails 95 and 97. This area has a large herd of mountain goats and in July they concentrate around the Black Mountain Lookout and are quite tame. You can come down to the river from Black Mountain on either Trail 396 or 399. Both are steep and hard on the knees. Throughout the area you may see elk, white-tailed deer, moose, or black bear. Forest birds you may see include the common raven, blue grouse, Clark's nutcracker, gray and Steller's jays, and mountain and chestnut-backed chickadees. A third trail, on the southern end of the pioneer area, runs for seven miles from one mile downstream of Aquarius Campground to Thrasher Creek. This trail offers a scenic trip through very old stands of cedar and hemlock forest similar to the Heritage Grove. The forest floor is covered with several kinds of ferns making the area very shaded and cool.

Directions: From Orofino head east for about 23 miles to Idaho 11 and turn north. In six miles you will come to the town of Headquarters. Bear left onto Forest Road 247 for about 21 miles to the North Fork Clearwater River bridge, which crosses the river at the Aquarius Campground. Turn left onto Forest Road 700 and go two miles. Bear right onto Forest Road 705 for 1.5 miles to the Isabella Creek Trail. To reach the Smith Ridge Trail go nine miles from the bridge, on Road 700. The trail paralleling the North Fork Clearwater that leads from Aquarius Campground downstream begins at Road 700. Clearwater National Forest maps are available at the North Fork District Ranger office in Orofino.

Ownership: USFS (476-3775, 245-4517)
Size: 30,000 acres   **Closest Town:** Headquarters

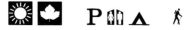

17 Hells Gate State Park

Description: This is a great birding area within easy access of Lewiston. The Snake River borders the park in a desert foothill setting. From the campground there is a two-mile trail that follows the river past basalt cliffs, prickly pear cactus, shrubby riparian vegetation, and grasslands. Of the park's 121 recorded bird species, northern oriole, eastern and western kingbird, swallows, and wrens are common while bald eagles are winter visitors. Among the uncommon species are Vaux's swift, marsh wren, Forster's and common tern, black-crowned night heron, Lincoln's sparrow, varied thrush, long-eared owl, and common loon. Mammals often spotted include Nutall's cottontail, white-tailed jackrabbit, and yellowbelly marmot while mule deer, river otter, mink, and raccoon are less common. Area outfitters offer one- and two-day jet boat trips up the river into Hells Canyon, the deepest gorge in North America. Boat trip arrangements can be made at the park and in Lewiston.

Directions: From the Interstate bridge at Lewiston, drive south along Snake River Avenue for four miles to the park entrance. A map and bird checklist are available at the visitor center. The park is open from March 1 to November 1.

Ownership: SP (743-2363), ACE
Size: 960 acres **Closest Town:** Lewiston

Spectacular scenery, such as Heart Lake, shown here, is one of the rewards of backcountry exploration in the Mallard-Larkins Pioneer Area. CRAIG GROVES

31

18	Craig Mountain Wildlife Management Area

Description: This large WMA begins at the bottom of Hells Canyon, along the Snake River, and climbs steeply upslope from 800 feet to one mile high. The varied elevations provide several habitat types. Flats along the river are covered with cheatgrass and sand dropseed with scattered hackberry trees, which attract spring songbirds. Upslope there are extensive fields of bluebunch wheat-grass and Idaho fescue attracting gray partridge and chukars. Wet draws full of brush and deciduous trees shelter turkeys, quail, and songbirds while middle and upper elevations give rise to conifers (mainly ponderosa pine and Douglas-fir) that harbor ruffed and blue grouse. The WMA is managed primarily for big game and upland birds and supports up to 1,000 mule deer, a few white-tailed deer (at higher elevations), 200 elk, and 30 bighorn sheep. Bighorn sheep were reintroduced in 1977 when several Washington-stocked sheep swam the Snake River and again in 1984 when the IDFG transplanted sheep from Wyoming. The easiest sheep viewing opportunity is from the road along the Washington side of the Snake River, from Asotin to the mouth of the Grande Ronde River. The higher elevation areas contain many snags, making great woodpecker habitat. White-headed woodpeckers, an uncommon species in Idaho, are fairly common here. Great gray owls have also been seen with increasing frequency on Craig Mountain. A few amphibians, reptiles (watch out for rattlesnakes), and numerous wildflowers and cactus provide secondary viewing opportunities. Hunting is popular from September through December. Motorized vehicles are prohibited in the WMA. Visitors may park at one of three parking areas at the perimeter, or access 13 miles of riverfront by boat. Three miles of riverfront are privately owned at the mouth of Captain John Creek. There is no developed trail system, but the country is very open and numerous game trails provide good hiking in most areas.

Directions: Enter Lewiston from U.S. 12, cross the Clearwater Memorial Bridge, and then head south on 21st Street. Follow the main road, which will change from 21st Street to Thain Road to 14th Street to Ripon to Tammany Creek Road. About 3.5 miles past Tammany School turn right onto Waha Road and go about seven miles to 21 Ranch (a white victorian house and barn). From this point follow the WMA signs. Turn right at the ranch and go 4.8 miles to Redbird access parking or 4.2 miles to Captain John access parking. To reach the third parking area continue south on Waha Road for about 16.5 miles, turn right at Madden Corral, and travel 3.9 miles. There are spectacular views along this route. Land access is closed when snow-covered, while river access is open all year. Power boaters may launch from ramps at Lewiston, Idaho, or in Washington at Clarkston, Asotin, or Heller Bar near the mouth of the Grande Ronde River. Maps are recommended and are available at the IDFG office at 1540 Warner Ave. in Lewiston. Toilets are along Snake River beaches. There are no developed campgrounds but camping is allowed.

Ownership: IDFG (743-6502), BLM (962-3245)
Size: 24,200 acres **Closest Town:** Lewiston

19   Winchester State Park

Description: At 4,000 feet in elevation, this lake at the eastern foot of Craig Mountain is surrounded by conifers and brushy hillsides, with a small marshy area at the lake inlet. Common birds are ruffed grouse, red-breasted and white-breasted nuthatches, common loon, evening grosbeak, and Steller's jay plus several waterfowl species. Deer and sometimes elk can be seen in early mornings and evenings from spring through fall. Some uncommonly seen species include osprey, turkey vulture, spotted sandpiper, pileated and white-headed woodpeckers, northern flying squirrel, long-tailed weasel, and coyote. Bald eagles, peregrine falcons, and northern goshawks are rare. There is a nature trail next to the park headquarters and other trails around the lake.

Directions: The park is in Winchester, one mile south of U.S. 95 (33 miles southeast of Lewiston). Check with the park headquarters for information on its interpretive programs and trails.

Ownership: SP (924-7563)
Size: 550 acres   **Closest Town:** Winchester

20   Nez Perce National Historical Park—Heart of the Monster Unit

Description: As legend has it, the "Heart of the Monster" is a basalt outcropping from which the Nez Perce Indian Tribe was created. It is situated next to the pristine Clearwater River and surrounded by steep mountains covered with pine and fir. For the best wildlife viewing, follow the paved foot trail beside the river. In the spring and summer look along the river for numerous nesting songbirds, mergansers, Canada geese, osprey, beaver, river otter, mink, and muskrat. Common songbirds include the yellow warbler, northern oriole, meadowlark, warbling vireo, common yellow-throat, fox sparrow, and rufous-sided towhee. White-tailed deer are often seen at dawn and dusk year-round. Elk may be present in winter as well as bald eagles, tundra swans, and many other waterfowl species. This is a popular fall and spring steelhead fishing spot.

Directions: The park is on U.S. 12, five miles north of Kooskia and two miles south of Kamiah.

Ownership: NPS (843-2261)
Size: 160 acres   **Closest Town:** Kamiah

| 21 | **Musselshell Meadows** |

Description: This site presents a rare mix of marshy meadow habitat within dense coniferous forest. The willow-lined Musselshell Creek flows through boggy meadows and past an old mill pond. The area has a diversity of wildlife, especially in the spring and summer. Moose with calves, elk, mule, and white-tailed deer, otter, numerous woodpecker species, and waterfowl are common. Follow the nature trail or walk through the area; move slowly and quietly to limit wildlife disturbance. Look for yellow warblers and vireos in the willows and northern phalaropes and American bitterns in the waterside vegetation. Listen for the call of common snipe and the honks of transitory snow geese on early spring mornings. This is one of the few areas in the state that supports fishers, a protected furbearer inhabiting mature coniferous forests and an IDFG species of special concern.

Directions: From Orofino, follow U.S. 12 south for eight miles to the junction with Idaho 11. Cross the Clearwater River and head east on Idaho 11 for approximately 16 miles to Weippe. Continue east through town on the paved road for eight miles to the intersection of Forest Road 100 and Peterson Corners. Turn right onto Road 100 for approximately four miles. Then turn left onto Forest Road 540 for 0.5 mile to the parking area. To reach the area from the south, turn right off U.S. 12 immediately before crossing the Clearwater River bridge at Kamiah. This is Road 100; follow for 25 miles to the meadows. Road 100 is maintained through the winter. There is a picnic area and pond adjacent to the parking lot and a nature trail with interpretive signs around the meadows. The meadow is encircled by roads, so you may look for wildlife from your car. Clearwater National Forest maps are available at Forest Service stations in Orofino, Kamiah, and Kooskia.

Ownership: USFS (476-4541)
Size: 100 acres **Closest Town:** Weippe

Musselshell Meadows and the Lochsa River Canyon are excellent places to see moose. Be sure to watch from a distance, though, because moose can be dangerous, especially cows with young calves. D. LEIGHTON

| 22 | **Lochsa River Canyon** |

Description: This scenic river canyon, part of the National Wild and Scenic River System, features steep slopes covered by a mosaic of evergreen trees and deciduous shrubs. There are frequent opportunities to stop and scan for a high diversity of wildlife, although only the most common animals are easily seen. The diverse vegetation created by large wildfires in the early and mid-1900's supports one of the nation's largest elk herds. Early mornings and late afternoons, especially in winter and spring, are good for viewing elk and white-tailed deer on the steep canyon slopes. An excellent place to view moose during the summer and fall is around the salt licks at the Elk Summit Cabin south of the Powell Ranger Station. Although a two-hour drive from the main highway, the moose viewing is well worth the drive. Wilderness Gateway Campground, located near mile post 122, 20 miles east of Lowell, has good birding trails. Common songbirds include red-eyed, solitary, and warbling vireos, yellow, yellow-rumped, orange-crowned, and McGillivray's warblers as well as hummingbirds, thrushes, and flycatchers. During late spring and summer listen for the calls of the belted kingfisher, Steller's jay, and Canada goose along the river, where osprey, common merganser, and harlequin ducks nest. Look for bald eagles and river otters in the winter. The Lochsa River and its tributaries provide important spawning and rearing habitat for the steelhead trout and chinook salmon. A 30-minute self-guided auto tour of USFS fishery habitat improvement work takes you a mile up Pappoose Creek Road near milepost 158. Many backcountry trails originate along U.S. 12. For shorter walks, trail bridges at mileposts 112, 136, and 152 lead to the south side of the river. Two short nature trails, Major Fenn (milepost 108) and Colgate Licks (milepost 148), interpret wildlife, fisheries, and other natural resources of the river corridor. At Lowell is the Three Rivers Resort restaurant, where several hummingbird feeders attract black-chinned, calliope, and rufous hummingbirds easily viewed while dining. Visitors passing through Montana may stop at the Lolo Pass visitor center for additional information on the Lewis and Clark expedition and area fish and wildlife viewing opportunities. Occasionally moose, elk, and deer are seen nearby. There are good mountain biking trails in the area, where you may see western tanager, Steller's and gray jay, Clark's nutcracker, mountain bluebird, pine siskin, and Swainson's and hermit thrush. Rafting is popular on the Lochsa River.

Directions: Take U.S. 12 to Lowell, 28 miles east of Kooskia, where the Lochsa and Selway Rivers join, and continue upstream for 65 miles to near the Powell Ranger Station. To reach Elk Summit, turn off U.S. 12 at the Powell Ranger Station sign near milepost 162. Follow this gravel road for two miles and turn left immediately after crossing the river (onto Forest Road 111). In three miles take the right fork (Forest Road 360) and continue another 12 miles to Elk Summit. Three Rivers Resort is adjacent to the Lochsa River just upstream from where it joins the Selway River. USFS maps are available at the Kooskia Ranger Station. Lolo Pass also has maps and information on the area.

Ownership: USFS (926-4275, 942-3113)
Size: 65 miles of river **Closest Town:** Lowell and Powell

23 Selway River Canyon

Description: The outstanding natural resources of the Selway River corridor earned its designation as a Wild and Scenic recreation river. The waters are clear and cold, and the corridor is aesthetically exceptional with steep, forested terrain and fern-lined trails. Only 20 miles of the river is accessible by road, from its junction with the Lochsa River (which then forms the Middle Fork of the Clearwater River) upstream to the Selway-Bitterroot Wilderness boundary. Wildlife is abundant throughout the corridor. Winter and spring are the best times to view elk, white-tailed deer, black bear, moose, bald and golden eagles, Canada geese, beaver, and river otter. In late spring to fall look for waterfowl (including the rare harlequin duck), osprey, and numerous songbirds. Johnson Bar, a 0.25-mile serene stretch of sand and gravel bar, is a good place to view waterfowl and deer. Cedar Flats is widely known as a white-tailed deer fawning area—a great opportunity to see and photograph deer in the early morning and evening hours from early to mid-summer. Maintain a good distance from the deer to avoid unnecessary disturbance. The 1.5-mile O'Hara Interpretive Trail follows the Selway for 0.5 mile then climbs into a cool cedar forest with lush ferns. There are wildlife information signs at the trailhead. Look for pileated woodpecker in winter and typical forest birds from spring to fall. Recreation opportunities include cross-country skiing at O'Hara and Johnson Bar and rafting and fishing along the river.

Directions: From Lowell, 28 miles east of Kooskia on U.S. 12, cross the river and continue up the Selway River Road. Johnson Bar is four miles from Lowell, Fenn Ranger Station (a good local information source) is a mile further, Cedar Flats wetland is another 0.5 mile up the road, and the O'Hara Interpretive Trail, which takes off from the east end of O'Hara Campground, is one mile further.

Ownership: USFS (926-4258)
Size: 20 miles of river **Closest Town:** Lowell

Cedar Flats Wetland along the Selway River provides opportunities for watching female white-tailed deer and their spotted fawns, such as the one shown here. MICHAEL H. FRANCIS

24 Elk City Area

Description: This highly scenic area includes the steep, forested canyons of the South Fork Clearwater River drainage and several of its crystal clear stream tributaries. Wildlife abounds in this remote and relatively undisturbed area, but the animals are secretive in their daily activities. You will be rewarded by spending some time on the area's abundant forest trails and back roads. There are six stops mentioned for this area. The first is at McAllister Campground. Here, a 1.25-mile interpreted trail leads to a viewing area where elk, white-tailed deer, and a few bird species can be observed in winter and early spring. Bald eagles winter along the South Fork Clearwater River. The second stop is at Meadow Creek Campground where a 0.25-mile foot trail leads to waterfalls with migrating steelhead trout that jump upstream to traditional spawning grounds from April through May. The third and fourth stops are at stream sites that have been rehabilitated following dredge mining damage to re-establish critical spawning and rearing sites for the Spring chinook salmon, which can be viewed from mid-August to mid-September. You may also see moose, elk, and white-tailed deer here from spring through fall. The fifth stop is at a large mountain meadow where from late-April through May up to 300 elk have been seen feeding on springtime meadow growth; wildflower blooms can be spectacular at this time. Wet portions of the meadow harbor waterfowl and other waterbirds, and raptors are commonly sighted perched in the trees bordering the meadow. View from your car, so as not to disturb the elk. The last stop is a chinook spawning, rearing, and holding facility that features guided tours in the summer and an opportunity to see fish up close.

Directions: From Grangeville, go northeast on Idaho 13 for approximately eight miles to the South Fork Clearwater River bridge where Idaho 13 and 14 join. Go south on Idaho 14 for about 10 miles to the McAllister Campground and Wildlife Viewing Trail, then five more miles to the Meadow Creek Campground and trail. Continue on Idaho 14 for approximately 19 miles and turn left onto Forest Road 858, which follows Newsome Creek. Drive alongside the creek for two miles where there will be turnouts in the road to park. Walk from the turnouts down to the creek. Return to Idaho 14 and continue east for about five miles. Make a right turn onto Forest Road 233, which follows Crooked River. Drive alongside Crooked River for up to six miles, parking wherever there are wide road turnouts, and walk down to the river. Return to Idaho 14 and then turn right onto Forest Road 222 after three miles. In eight miles you will come to a large meadow on your right. Watch for the road turnouts within the next two miles and use them to view elk and other wildlife. In another two miles you will come to the Red River Ranger Station, which houses the IDFG chinook salmon facility. There are campgrounds along Crooked and Red rivers. A Nez Perce National Forest map is recommended for this tour and is available in Grangeville.

Ownership: BLM (962-3245), USFS (983-1950), IDFG, PVT
Size: 45 miles one-way **Closest Town:** Elk City

25 | Lower Salmon River Canyon

Description: The Salmon River Canyon is one of the nation's deepest gorges—second only to Hells Canyon—and the nation's longest undammed river. The river is bordered by canyon grasslands and steep rocky cliffs, and is unique because of its Upper Colombian River Basin flora and fauna. The best way to experience the area is by a commercial or self-guided float trip. Common wildlife include the golden eagle, red-tailed hawk, prairie falcon, American kestrel, mule deer, white-tailed deer, elk, Canada goose, chukar, and river otter. Although most animals can be seen year-round, the river is not floatable during the icy winter months. Hoofed mammals are best seen during winter and spring. Self-guided boaters (recommended only for experienced whitewater travelers) can launch at Riggins, Lucile, Slate Creek, Twin Bridges, Hammer Creek, and Pine Bar. Those viewing by car should be sure to use turnouts when looking for wildlife. Several commercial guide companies are located in Riggins; they offer half-day to eight-day river trips (contact the Riggins Chamber of Commerce, 628-3456).

Directions: U.S. 95 parallels the river from Riggins downstream to White Bird. Approximately 14 miles further downstream (northwest) the river is accessed from Cottonwood by driving 10 miles south on the gravelled Graves Creek Road. Large portions of the river are accessible only by jet boats or float boats. Commercial guides normally take out at Heller's Bar at the mouth of the Grande Ronde River, Washington, or at Lewiston. The BLM has produced a guide to the Lower Salmon River, which is available at their Cottonwood office. Float and power boat permits are required June 20 to September 7 for those venturing below Hammer Creek and are available at the FS Fire Center in Riggins. Permits and Nez Perce Forest maps are available at the Slate Creek Ranger Station. There are day-use areas at Slate, Skookumchuck, and Hammer Creeks.

Ownership: BLM (962-3245), PVT
Size: 87 miles of river **Closest Town:** Riggins, White Bird

Though appearing rodent-like, the tiny pika is actually related to rabbits. It inhabits high elevation talus slopes throughout Idaho. DENNIS AND MARIA HENRY

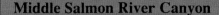

26 | Middle Salmon River Canyon

Description: The "River of No Return," as dubbed by the Shoshoni Indians, roars through the pristine Frank Church River of No Return Wilderness, the largest single wilderness in the lower 48 states. You will travel through a plunging river gorge with magnificent canyon walls and deeply dissected side canyons. River floaters can often spot mountain goats and bighorn sheep, which can be seen year-round in small groups but are best viewed in spring. Also look for golden eagles, waterfowl, and river otters. The adjoining wilderness area is home to elk, mule and white-tailed deer, moose, black bear, mountain lion, and hundreds of other wildlife species. The numerous hiking trails range from easy to very difficult. The river runs spring through fall, until ice forms. Contact the USFS for maps and assistance in planning a backcountry trip.
Additional viewing area: The Middle Fork of the Salmon River, within the Frank Church River of No Return Wilderness, offers 100 miles of outstanding whitewater rafting and good wildlife viewing opportunities. Contact the Middle Fork Ranger District, 879-4321, to plan a trip.

Directions: For downstream access, take U.S. 95 to Riggins. At the south end of town look for a bridge that crosses the Little Salmon River. This will lead to Forest Road 1614 (gravel), which borders the Salmon River for 25 miles. The road will end at Vinegar Creek, where a jet boat is needed to continue upstream. You can also float from here back to Riggins. Several commercial guide companies in Riggins offer half-day to eight-day river trips (contact the Riggins Chamber of Commerce, 628-3456). For upstream access, drive north of Salmon on U.S. 93 to North Fork. From here take Forest Road 30 west to the end of the road at Corn Creek (44 miles on gravel road). Rafts and kayaks can be launched in the water at any point along the river outside of North Fork. Commercial guide trips are available through many companies in Salmon (contact the Salmon Chamber of Commerce, 756-2100). Unguided trips are recommended only for experienced whitewater travelers; permits are required for trips below Corn Creek.

Ownership: USFS (839-2211 Slate Creek, 865-2383 North Fork)
Size: 80 miles of river **Closest Town:** Riggins, North Fork

To assure the continuous protection of free-flowing waterways, Congress created the National Wild and Scenic River System. Idaho boasts five rivers that are a part of this system—the Snake, Salmon, Selway, Rapid, and Middle Fork Salmon.

Experts at living on the edge, mountain goats impress wildlife watchers with their fearless rock climbing abilities. Great sites to see them include Farragut State Park in North Idaho and the Seven Devils Mountains in North Central Idaho. TIM CHRISTIE

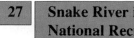
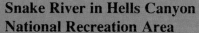

| 27 | **Snake River in Hells Canyon National Recreation Area** |

Description: The Snake River, from Hells Canyon Dam to the Oregon/Washington state line, bisects the Hells Canyon Wilderness and travels through the deepest canyon in North America, with rock cliffs up to 5,000 feet tall and gentler grassland slopes. This river section is designated "Wild" from the dam to Pittsburg Landing (31.5 miles) and "Scenic" from the landing to the NRA's northern boundary (36 miles). The rugged terrain and isolation provides habitat for many uncommon species including Townsend's big-eared bat, MacFarlanes' four-o'clock, and many rare insects. Wildlife viewing is year-round by boat or foot, with best viewing opportunities below Pittsburg Landing. Most people see the area by a commercial raft trip; animals generally let you approach closer by boat. Common species include chukar, great blue heron, elk, mule deer, bighorn sheep, and mountain goat. Occasionally you may see golden eagles, Canada geese, and black bear (spring). Bald eagles, elk, mule deer, bighorn sheep, and mountain goats concentrate near the river in winter. On the 12-mile stretch from Duck Bar to Jim Creek, below Pittsburg Landing, bighorn sheep are often seen. Foot trails parallel both sides of the river. The Idaho trail starts as a path near Brush Creek (accessible by jet boat shuttle), improves to an all purpose trail (mountain bikes allowed) at Granite Creek, and ends at Pittsburg Landing. You can hike in several miles from the Seven Devils Mountains area to the trailhead or take a two-mile trail at the end of Road 242, near Kirkwood Creek. If traveling overland be aware of rattlesnakes and poison ivy. Commercial outfitters shuttle people to Hells Canyon Dam from Pine Creek, Oregon, near Oxbow Dam and will also bring them back to Hells Canyon Dam from Pittsburg Landing (a two- to four-day float trip). There are also jet boat tours from the base of the dam, Pittsburg Landing, and Lewiston.

Directions: For upstream access (all paved roads), take U.S. 95 to Cambridge, then take Idaho 71 northwest for 29 miles to Oxbow Reservoir. Idaho 71 ends near the Oxbow dam. Cross the river and continue on Forest Road 454 along Hells Canyon Reservoir to the end of the road at Hells Canyon Dam. To reach Pittsburg Landing, take U.S. 95 to White Bird, then head west on Road 493 for 17 miles. The roads are gravel and dirt; passenger cars are not recommended when the roads are wet. To access the riverside hiking trail upstream of Pittsburgh Landing, take U.S. 95 to Lucile, then take Road 242 west for about 12 miles. This road is rugged and somewhat hard to find. Permits are required for all boating from the Friday preceding Memorial Day weekend through September 15. Contact the Hells Canyon NRA in Lewiston for a map of the area, backcountry rules, and outfitter information. There is also a HCNRA office in Riggins.

Ownership: USFS (743-2297)
Size: 67.5 miles of river **Closest Town:** Oxbow (Oregon), White Bird

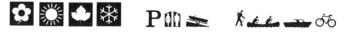

28 Seven Devils Mountains in Hells Canyon National Recreation Area

Description: These seven rocky alpine peaks climb skyward from the Snake River to over 9,300 feet in elevation and are often snow-capped into July. They provide excellent habitat for mountain goats—the main wildlife attraction at this site—which are best viewed from July to September. Heavens Gate Lookout offers an incredible view of portions of Washington, Oregon, Idaho, and Montana. Several hiking trails lead to the over 30 alpine lakes. Watch for golden eagles, yellow-bellied marmots, Columbian ground squirrels, pika, and the tracks of black bear and coyote. Seven Devils Lake is a popular canoeing spot. There are spectacular meadow wildflower blooms in July with dogtooth violet, trillium, Rocky Mountain iris, and wild hyacinth, to name a few. On the drive up watch for elk, white-tailed deer, and ruffed and blue grouse.

Directions: From Riggins, head south on U.S. 95 for 0.75 mile. Turn right onto Forest Road 517 and go 17 miles to the Seven Devils Campground (just west of Windy Saddle Campground). From the parking lot scan the cliffs, 0.25 mile to the southwest, for mountain goats. Heavens Gate Lookout is 1.25 miles further down Road 517. Maps of the area are available at the USFS office, which is well signed from the highway to the south of Riggins.

Ownership: USFS (628-3916)
Size: Over 50 square miles **Closest Town:** Riggins

29 Rapid River Fish Hatchery

Description: This fish hatchery is located along the Wild and Scenic Rapid River. The steep slopes of the river canyon are covered with open stands of ponderosa pine and Douglas-fir, intermixed with stands of mountain mahogany, greenbush, and birch. The hatchery offers a close look at chinook salmon and a lesson on their life cycle. View adult chinook from May to mid-September, and juveniles year-round. The hatchery is open all year, seven days a week, with tours available anytime. A Forest Service trail bordering the river, upstream of the hatchery, is a good place to see wrens, kinglets, Townsend's solitaires, golden eagles, northern goshawks, chukar, and gray partridge. The chestnut-backed chickadee has been seen here during the winter. Watch for western rattlesnakes along the trail, especially in low water years.

Directions: From Riggins, take U.S. 95 south for 4.4 miles, then turn right onto Forest Road 2114 for 2.7 miles.

Ownership: Idaho Power Company, managed by IDFG (628-3277)
Size: Five acres **Closest Town:** Riggins

SITE	SITE NAME
30	Brownlee, Oxbow, and Hells Canyon Reservoirs
31	Ponderosa State Park
32	Cascade Reservoir
33	Bear Valley
34	Montour Wildlife Management and Recreation Area
35	Fort Boise Wildlife Management Area
36	Owyhee Mountains
37	Deer Flat National Wildlife Refuge
38	World Center for Birds of Prey
	Boise River Greenbelt
39	Barber Park
40	Boise Nature Center
41	Kathryn Albertson Park
42	Boise River Wildlife Management Area
43	Upper Black's Creek
44	Snake River Birds of Prey Area
45	Ted Trueblood Wildlife Area
46	C. J. Strike Wildlife Management Area
47	Bruneau Dunes State Park

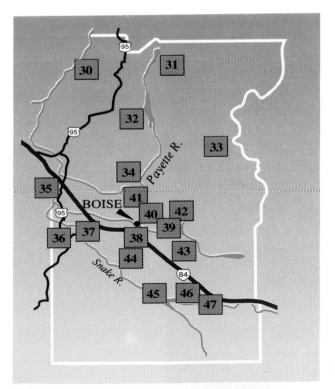

30 | Brownlee, Oxbow, and Hells Canyon Reservoirs

Description: The Snake River flows for 70 miles through three reservoirs before dropping into Hells Canyon, the deepest canyon in North America. The reservoirs are flanked by sagebrush and bitterbrush shrub steppe and surrounded by scattered ponderosa pine and Douglas-fir trees. This remote, uncrowded area is an important wintering range for mule deer, elk, and bald eagles, with up to 50 eagles seen at one time on Oxbow Reservoir alone. One can see deer and elk right along the roadway from December to March, when many waterfowl species use the reservoirs. River otter and upland birds, especially chukar, are also common in the area, although not frequently seen. Bighorn sheep can be spotted near Hells Canyon Dam. A stairway at the dam allows foot access to the base and to a trail on Deep Creek. (For information on exploring below the dam see site No. 27 in Region II.)

Directions: Take U.S. 95 to Cambridge, then take Idaho 71 northwest for 20 miles. You will first come to Brownlee Reservoir. Boat access to Brownlee is best from Woodhead Park. Note the turnout just after Brownlee comes into view. To see the reservoir by car take the west shoreline road, just after crossing over the Snake River below the dam. Continuing on Idaho 71, drive nine miles along the west shoreline of Oxbow Reservoir. Idaho 71 ends near the Oxbow dam. Cross the river again and continue on Forest Road 454 along Hells Canyon Reservoir to the end of the road at Hells Canyon Dam.

Ownership: BLM (334-1582), USFS (628-6916), Idaho Power Co.
Size: 70 miles of river **Closest Town:** Cambridge

Flickers are cavity nesters, excavating tree cavities that are often later used by reptiles, birds, and small mammals. These woodpeckers get their name from their undulating flight pattern. JIM PETEK

Description: This state park, named for its very old ponderosa pines, sits on a two-mile forested peninsula that almost bisects Payette Lake. The park has varied topography and habitats, with arid sagebrush flats, steep basaltic cliffs, dense conifer groves, meadows, and marshes. The park offers self-guided nature trails, guided walks with park naturalists, and evening campfire programs. Of particular interest are two marshes evolving to meadowland. Wildlife diversity is high, although many species can only be seen at dawn or dusk. During the spring and summer months look for Pacific tree frogs, western toads, long-toed salamanders, and garter snakes in the two marsh areas as well as mule deer, grouse, great horned and barred owl (very rarely the great gray owl), pileated woodpecker, Swainson's thrush, and many other songbirds. Black bear, red fox, badger, and bobcat are rarely seen, although tracks left during nocturnal visits may be spotted. Common winter wildlife include red squirrels, Nutall's cottontail, red fox, mink, woodpeckers, nuthatches, osprey, and bald eagles. The park is also rich in floral diversity with bright wildflower blooms in spring and summer. Look for lilies, penstemmon, spring beauties, trillium, and clematis. For additional viewing, the McCall Fish Hatchery raises Summer chinook, a threatened species. Visitors can observe chinook rearing from the egg stage to the pre-smolt stage. The best time to visit is May through December. There is a visitor center and a self-guided tour.

Directions: Drive north through McCall on Idaho 55 and take the right fork at the downtown "Y" intersection. In two miles you will see the park entrance on the left. The visitor center has park and trail maps, bird and plant checklists, and wildlife notes available. The McCall Hatchery is on the north side of Idaho 55 (on the south end of town) and is open year-round.

Ownership: SP (634-2164)
Size: 900 acres **Closest Town:** McCall

Red fox pups play at their den. Red foxes are at home in farmlands, sagebrush, grasslands, and open forests throughout Idaho. JAN L. WASSINK

Description: Cascade Reservoir is rimmed on the east by open fields and on the west by coniferous forest. Water birds are the main attraction here. When the reservoir is full in summer it is best to observe from the road; when it is drawn down in the spring, fall, and winter you can walk the shoreline. The mouth of Duck Creek is a major nesting area for western and Clark's grebes; bald eagles and osprey also nest at the reservoir. The calls of common loons are frequently heard. Mudflats around the reservoir are good for viewing migrating shorebirds and waterfowl. On the east side, the numerous small streams and ponds that dot the mudflats provide a vast, open area for water birds until the reservoir fills. Forested areas are good for seeing Lewis', downy, hairy, pileated, and black-backed woodpeckers as well as the red-naped sapsucker. Great gray and barred owls have been observed in the lodgepole pine forests and peregrine falcons have been sighted at the reservoir. Uncommon birds include Vaux's swift, varied thrush, mountain bluebird, Townsend's warbler, American redstart, water pipit, white pelican, and sandhill crane. Winter is poor for wildlife viewing, although you may see hooded merganser, and very rarely oldsquaw, below the bridge on the north end of Cascade. For additional viewing, continue east of Cascade Reservoir to the South Fork Salmon River drainage where chinook salmon spawn each August, migrating over 700 miles from the ocean to the upper reaches of the river and its tributaries. Stolle Meadows viewing site has a boardwalk with interpretive signs and a great view of the fish spawning. Downstream of the meadows are two viewing sites where fish can be seen pairing up to spawn; the best viewing is in mid-August.

Directions: Take Idaho 55 to Donnelly, then turn west onto a paved road and drive four miles. At the "T" intersection, turn left (south) onto Forest Road 422. This gravel road follows the west shoreline of the reservoir for the next 16 miles. The road will then turn to pavement and, in six miles, lead you back to Idaho 55 at Cascade. There are 12 public campgrounds along the reservoir for good shoreline access. To reach the east side of the reservoir, follow Idaho 55 until 0.5 mile south of Donnelly, turn right, and proceed due south across Goldfork Creek (where every other transmission tower holds an osprey nest). In another three miles the road enters the mudflats where road maintenance stops and walking is recommended. The USBR has a free brochure of Cascade Reservoir. Forest Service maps of the area can be purchased at the Cascade office. To reach the salmon viewing areas, turn east onto Warm Lake Road, just north of Cascade, and go about 20 miles to the South Fork Salmon River. Cross the river, turn right onto Forest Road 474, and go four miles to Stolle Meadows. You can also view the fish at about 0.5 mile and at 15 miles (Poverty Flat) downstream of Warm Lake Road, on Road 474. This road eventually loops back to McCall at Idaho 55.

Ownership: USFS (382-4271), USBR (382-4258)
Size: 28,300 acres **Closest Town:** Donnelly, Cascade

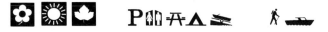

Natural fires play an important role in maintaining habitat diversity. Many animals require specific types and age stages of habitats to survive.

33 | Bear Valley

Description: This portion of the Boise National Forest is an excellent example of central Idaho's high mountain country, with many mountain meadows surrounded by stands of lodgepole pine and Douglas-fir. From July to September, in the early morning and evening hours, mule deer and elk come out of the timbered forests to feed in the meadows. Observe them from your vehicle. Common bird species include sandhill crane, great blue heron, northern harrier, Swainson's hawk, American kestrel, mountain bluebird, gray jay, and belted kingfisher. Great gray owls, the largest owl in Idaho, have also been observed in the valley. Several waterfowl species can be seen in the ponds at the upper end of Bear Valley Creek. During July and August see spawning salmon in the creek and observe the mountain meadow wildflower blooms. The tracks of the elusive wolverine have sometimes been seen in winter here. Whitehawk Mountain Fire Lookout provides a wonderful view of central Idaho. From the intersection of Forest Roads 510 and 582, drive 3.5 miles north on 582 to Forest Road 569 and follow to the lookout. When the road is muddy or snowy, only 4-wheel drive vehicles are recommended.

Directions: Take Idaho 21 to Lowman, then take Forest Road 582 (gravel) north for 15 miles to the headwaters of Bear Valley Creek. Follow 582 through the valley and to Bruce Meadows (the road will become 579) and back to Idaho 21 near Cape Horn. From this point you can return to Lowman via Idaho 21 (Ponderosa Scenic Route) or head east eight miles to the Sawtooth National Recreation Area. There are many hiking trails in the area and a Forest Service map, available at their Lowman office, is recommended. There is a rest area at Bruce Meadows.

Ownership: USFS (259-3361)
Size: 16-mile long valley **Closest Town:** Lowman

Osprey abound at Cascade Reservoir. They are easily observed fishing and tending to breeding activities on artificial platforms and treetop nests. GLENN OAKLEY

47

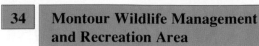

34   Montour Wildlife Management and Recreation Area

Description: This WMRA's flooded fields in the Payette River Canyon make excellent habitat for water birds. In spring and early summer the area is filled with Canada geese, ducks, common snipe, and blackbirds, plus osprey, northern harrier, golden eagle, American avocet, Wilson's phalarope, northern oriole, American goldfinch, western tanager, western kingbird, and other flycatchers. Look for great blue heron nests downstream of the Payette River bridge and for common yellowthroat and yellow-headed blackbirds in the large cattail marsh on the area's south side. Virginia rail, in the cattails, will usually respond to taped calls. Great horned owls and red-tailed hawks often nest in the large trees on the WMRA's eastern side. Look for mule deer in the winter. Portions of the area are closed during the spring nesting season. For a scenic, easy, canoe trip put in at Horseshoe Bend and take out at Montour Bridge.

Directions: From Emmett, go 13 miles east on Idaho 52 (or from Horseshoe Bend, go eight miles west on Idaho 52). Turn south at the Sweet-Ola junction. In a mile you will cross the Payette River, which forms the northern boundary of the WMRA.

Ownership: USBR (334-1060), managed jointly with IDFG (327-7025)
Size: 1,055 acres   **Closest Town:** Emmett, Horseshoe Bend

35   Fort Boise Wildlife Management Area

Description: Fort Boise is situated at the confluence of the Snake, Boise, and Owyhee Rivers. Cottonwoods and willows line the waterways, with wetlands and artificial impoundments surrounded by cattail, bulrush, and sedges. Common species include white-tailed deer, turkey, ring-necked pheasant, American avocet, black-necked stilt, black-crowned night heron, great blue heron, great egret, western screech owl, Swainson's hawk, common yellowthroat, yellow-breasted chat, and yellow-headed blackbird. During the winter, bald eagles, merlin, and Cooper's hawks may also be seen. Harris' sparrow, yellow-billed cuckoo, winter wren, pectoral sandpiper, and white-faced ibis are rare. March and April are best for seeing migrating birds including the white-fronted goose. Fishing and hunting are popular here.

Directions: From Caldwell, take Interstate 84 exit 26 to Parma. From Parma, travel north on U.S. 20-26-95 for three miles then turn left (west) onto Old Fort Boise Road to the WMA headquarters (two miles). Pick up a map and area guide at the headquarters. Marsh areas are closed during the nesting season.

Ownership: IDFG (722-5888)
Size: 1,500 acres   **Closest Town:** Parma

36 | Owyhee Mountains

Description: The Owyhee's are a rugged and remote desert mountain range in the southwestern corner of Idaho. This site is best explored as a two-day trip, from May to October, but can be completed in one day. Viewing is from the road only for the entire route. A high-clearance vehicle is recommended. From Marsing to Jordan Valley you can occasionally spot pronghorn in the sagebrush grasslands and often will see northern harriers, ferruginous and red-tailed hawks, and golden eagles. Rough-legged hawks are common winter visitors. In Jordan Valley you will head east on the Jordan Creek Road. For the first four miles watch for ring-necked pheasant, sandhill cranes, and white-faced ibis (spring and fall), and mule deer and coyotes in the private meadow along Jordan Creek. If the road is open, sage grouse can be observed on their leks (traditional display grounds) from late February through mid-May from Goose Creek to Triangle Reservoir and from one mile north of Hyde Saddle to three miles south of Oreana. Viewing is only for the first few hours after daybreak. Endemic redband trout inhabit all perennial streams in the area and can be seen at the Duck, Jordan, Flint, North Boulder Creek, and Rock Creek road crossings. These streams also support riparian vegetation critical to the survival of desert wildlife. River otter, beaver, mink, bobcat, red fox, mule deer, upland birds, songbirds, amphibians, and reptiles all depend on riparian systems for food, water, and shelter. Just after turning north at Triangle Junction, stop by Spencer Reservoir (adjacent to the road) to view ducks, Canada geese, snow geese, tundra swans, grebes, and various shorebirds. Spring and fall are the best times for viewing. Hundreds of bluebird nesting boxes have been erected throughout the Owyhee Mountains by private parties. Several can be seen north of Hyde Saddle along the main road. From Oreana to Walters Ferry you will often see a diversity of raptors and can occasionally spot burrowing owls, long-billed curlews, pronghorn, and mule deer. At Walters Ferry, Marsing, and two sportsman's access points in-between, the road crosses or meets the Snake River; these are good places to look for waterfowl in the spring and fall and bald eagles in the winter.

Directions: An area map is recommended and available from the BLM Boise District Office. This 140-mile round-trip drive can be accessed from several directions but will be described as starting and ending in Marsing. From Marsing on Idaho 55 head west two miles and turn left (south) onto U.S. 95 for 48 miles to Jordan Valley. Turn left at the fork in town and at 4.5 miles out take the left fork (County Road 3706). Continue approximately 25 miles to Triangle Junction, turn left and continue over Toy Pass to Oreana (about 20 miles). The county roads are not maintained in the winter. Go north from Oreana one mile and head northwest on Idaho 78. You will pass Walters Ferry in about 23 miles and return to Marsing in another 20 miles. For winter to spring visits the entire route will not be passable. Length of entry from the east or west onto the county roads between Oreana and Jordan Valley will depend on the amount of snow cover. Carry drinking water and be sure you have a full tank of gas before embarking on this trip. Camping and mountain biking are permissible on public land.

Ownership: BLM (334-1582), IDL, PVT
Size: 140-mile drive **Closest Town:** Marsing, Jordan Valley (Oregon), Oreana

Description: Lake Lowell is a great birding spot with over 200 species recorded. Spectacular bird concentrations occur on the lake during peak migration periods. Shorebirds appear in large numbers in August when low water levels expose mudflats. Mallard, pintail, American widgeon, green-winged teal, wood duck, and Canada geese are numerous from September to December in the planted refuge fields. Visiting fall-winter raptors attracted by the abundance of avian prey include bald and golden eagles, northern goshawks, Cooper's, sharp-shinned, and rough-legged hawks, and prairie and peregrine falcons. During April and May watch for migrating songbirds. In spring and summer look for nesting western and Clark's grebes, double-crested cormorants, Caspian terns, soras, Virginia rails, and great horned, northern saw-whet, western screech, long-eared, and barn owls. The 113-mile long Snake River sector, with 107 Canada geese nesting islands, can only be viewed by boat. The islands are closed to public access from February through May. Check in at the refuge headquarters for more information.

Directions: *From downtown Nampa (just south of Interstate 84 and accessible by Idaho 55 or U.S. 30), take 12th Street south to Lake Lowell Avenue (a right turn) and head west for four miles to the refuge headquarters. The lake is closed to boating from October 1 through April 14.*

Ownership: USBR (334-6117), managed by USFWS (467-9278)
Size: 11,430 acres **Closest Town:** Nampa

| 38 | **World Center for Birds of Prey** |

Description: Overlooking southwest Idaho's Treasure Valley, the Center offers visitors an intimate view of predatory bird biology, research, and management. The 90-minute guided tour includes video and slide shows as well as viewing of bird breeding facilities. In addition to a large, successful captive propagation and release project for the endangered peregrine falcon, the center is home to conservation projects for many of the world's rarest birds of prey. During the tour you are likely to experience close-up views of peregrine and orange-breasted falcons and golden and harpy eagles. From April to June visitors may see incubating falcon eggs and nestlings.

Directions: *From Interstate 84, take the Overland-Cole Road exit 50 (Boise) and travel south on South Cole Road for 4.3 miles. The road then turns into Flying Hawk Lane (dirt). Follow the road for one mile to the parking lot. Tours are by reservation only.*

Ownership: The Peregrine Fund, Inc. (362-3716)
Size: 280 acres **Closest Town:** Boise

The American avocet feeds by rapidly sweeping its head from side to side in soft mud or muddy water as it searches for insects and small crustaceans. It may be seen at many wetland sites, including Deer Flat National Wildlife Refuge. DENNIS HENRY

Boise River Greenbelt

The Boise greenbelt pathway system links over 850 acres of parks and natural areas with the Boise River's riparian habitat. There is a 14-mile long, 10-foot wide bicycle and pedestrian path that is paved except for a 1.5-mile section from Barber Park downstream. Year-round visitors can peer through numerous native plant species to view songbirds, ducks, Canada geese, and great blue heron as well as bald eagles in the winter. Beaver, gray fox, and muskrat are also present, although rarely spotted. The river is popular for summer floating by innertube, raft, kayak, and canoe; put in at Barber Park and float six miles to Ann Morrison Park next to Kathryn Albertson Park. Wildlife viewing is best in the early morning and late afternoon hours. The next three sites are developed wildlife viewing sites adjacent to the greenbelt. Two have unique educational features and the other offers natural viewing close to an urban area. A map showing the greenbelt, parks, museums, and other attractions is available from the Boise Park System at 1104 Royal Blvd., Boise.

39   Barber Park

Description: This county park is a great place to view wintering bald eagles from mid-November through mid-March (peak time is January and February). Adult and immature eagles can often be seen perched in large cottonwood trees watching the water for fish. It takes four to five years for adult bald eagles to develop a white head; one-year old eagles are completely dark brown while older birds are mottled white and brown. Scan the trees across from the raft launch site or walk downstream one or two miles along the greenbelt trails. Stay at least 200 feet from the birds and view with binoculars or spotting scopes. If you must pass closer maintain a steady pace to prevent them from flushing. Common resident wildlife include great blue herons, mallards, American widgeon, downy and hairy woodpeckers, northern flickers, belted kingfisher, California quail, mink, muskrats, and red fox. Look along the trail for the tell-tale tree gnawings of beaver.

Directions: From Interstate 84, take the Broadway exit (Boise), turning north at the first light. In about one block, turn right onto Federal Way, go 0.5 mile, and turn left onto Amity Road. Drive two miles then turn left onto Healey Road for 0.5 mile. The park office is open from Memorial Day weekend through Labor Day weekend. There are innertube and raft rentals and a launching area.

Ownership: Ada County (343-1328)
Size: 40 acres **Closest Town:** Boise

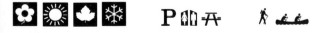

To bring wildlife closer to home, put up a nest box. Screech and saw-whet owls, bluebirds, finches, sparrows, and wood ducks readily take to boxes placed in the proper habitats.

40   Boise Nature Center

Description: With a river observatory and a wildlife interpretive area, the Center is an excellent place to learn about wildlife and fisheries issues and enhancement opportunities. The observatory is a 500-foot long, cold water stream replicating an Idaho river. Paths follow its meandering course, then dip below ground level at four underwater viewing stations. Windows reveal the components of an aquatic ecosystem and the life cycle of fish, from incubating eggs to adults. The interpretive area surrounding the stream presents visitors with a cross section of Idaho outdoors and includes five southwest Idaho plant communities, a formal backyard setting, a naturalistic backyard, and a rural farm plot and orchard. Signs present information on wildlife-related topics including habitat enhancement, wetlands, snag cover, and butterfly gardens. You will see bird feeders, nest boxes, and nesting platforms for various bird species. The Center's goal is to encourage an active, hands-on involvement of Idaho's citizens in wildlife management and habitat enhancement. Visitors will see numerous examples of what can be done to protect and preserve our natural resources. You can walk unguided through the area or schedule a guided tour.

Directions: From Interstate 84, exit onto Broadway (Boise) and drive north about three miles. Just after crossing the Boise River turn right onto Park Blvd. and go 0.3 mile. Turn right onto South Walnut Street and immediately turn left into the Idaho Department of Fish and Game parking lot.

Ownership: IDFG (334-2676) or (334-2675, messages)
Size: Four acres   **Closest Town:** Boise

41   Kathryn Albertson Park

Description: This walk-through park was designed as a place to view wildlife— particularly birds. A paved path meanders past ponds, trees, shrubs, and open lawn. The park attracts wildlife supported by the adjacent Boise River. Most conspicuous are waterfowl and songbirds. Two stout, beam and stone gazebos offer attractive views and seating along the paths. Several signs address a variety of wildlife issues including species identification and adaptations, endangered species, and the value of snags, wetlands, and riparian areas. This site also displays a portion of the world's largest known ponderosa pine tree. Guided tours are available by appointment with the IDFG.

Directions: Heading east on Interstate 84 Business Route to downtown Boise, exit right onto 16th St., which will soon turn into Americana Blvd. The park is located at 1000 Americana Blvd. just after you cross the Boise River.

Ownership: Boise City Parks, tours by IDFG (334-2676) or (334-2675, messages)
Size: 40 acres   **Closest Town:** Boise

42 | Boise River Wildlife Management Area

Description: The foothills of this WMA are covered by sagebrush-bitterbrush steppe, providing critical winter range habitat for mule deer. Over 6,000 deer winter on the slopes and can be observed from the main highway between December and March. Golden and bald eagles can be spotted along the Boise River and Lucky Peak Reservoir. During winter at sundown, rosy finches often come in to roost in cliff swallow nests on the cliffs across the highway from the Discovery Park entrance. Reservoir boat ramps can be accessed across Lucky Peak Dam at Turner Gulch or by crossing the reservoir and turning right (east) onto Forest Road 268 for four miles.

Directions: From Boise, take Idaho 21 east for four miles to the USFS Boise District Ranger office. Viewing begins here and continues along Idaho 21 for the next 16 miles to the upper reaches of Lucky Peak Reservoir. Use existing turnouts for wildlife viewing and do not approach deer on foot. Discovery Park is about 3.5 miles beyond the District Ranger office.

Ownership: IDFG (327-7025), BLM, ACE, IDL, USFS, PVT
Size: 16,650 acres **Closest Town:** Boise

Over 6,000 mule deer winter in the foothills east of Boise and are easily observed from State Highway 21. TIM CHRISTIE

| 43 | Upper Black's Creek |

Description: The riparian shrubs and trees lining Upper Black's Creek provide a haven for birds in this sagebrush desert setting. Viewing is at its peak during mid-May to early July and during fall migration. The first part of the tour travels through sagebrush and grassland desert country. Common bird species include the red-tailed hawk, horned lark, lark sparrow, and western meadowlark. About 4.5 miles from the freeway, just before a cattleguard, stop and look for western kingbirds and northern orioles in the locust trees below the road by the old corral. A few hundred yards farther up the road on the left is a small pond that usually has a family of ducks. In the trees around the pond look for downy woodpeckers, kinglets, warblers, and northern orioles. After the next hill are a string of willows where long-eared owls have nested in the past. Turning left at Mayfield Junction leads you to Black's Creek Road. About a mile after the junction the road bends to the right and there is a large grove of trees and shrubs. This is a good place to get out and walk as more birds are usually concentrated here than any other place on the creek. A partial list would be California quail, willow flycatcher, sage thrasher, yellow-breasted chat, black-headed grosbeak, lazuli bunting, and rufous-sided towhee. Birding habitat continues to be good all the way to the summit of the creek drainage.

Additional viewing area: If you are visiting the area between mid-August and mid-October and would like to view a large run of spawning kokanee salmon, continue on the Black Creek Road to the South Fork Boise River above Anderson Ranch Reservoir. It is about 45 miles from Black's Creek. The best viewing is below the bridge at Pine and at Baumgarter campground. On the way watch for small boxes mounted on trees, poles, and fence posts. These are bluebird nest boxes which are used by western and mountain bluebirds, tree swallows, and house wrens.

Directions: *Head east from Boise on Interstate 84, take Exit 64, turn left under the freeway, and proceed on the main traveled road. In approximately six miles bear left at Mayfield Junction. There is excellent bird watching for 4.5 miles, from here to the summit. From the summit you can continue on Road 189 for 22 miles to Prairie. Three-fourths of a mile further make a right turn onto Forest Road 128 and go approximately 23 miles to Pine where the road crosses the South Fork Boise River. View fish below this bridge and at Baumgarter campground. To reach Baumgarter head north from Pine on Forest Road FH61 nine miles to Featherville. Turn right onto Forest Road 227 at Featherville and go 10 miles to the campground. You can return to Interstate 84 via U.S. 20 by turning south at Pine onto FH61 for 15 miles to Idaho 20. Heading west on U.S. 20 will lead to Interstate 84 in 35 miles. You may want to stop by Little Camas Reservoir (in about eight miles) which has many riparian songbirds, bittern, and Virginia rail. Heading east, you will come to another wildlife viewing site, Camas Prairie Centennial Marsh (see Region VII). Boise National Forest maps are available from the Forest Service offices in Boise and Mountain Home. There are campgrounds, gas, and food near Pine. There are no facilities at Upper Black's Creek.*

Ownership: PVT, USFS (334-1516)
Size: Five miles of creek **Closest Town:** Boise

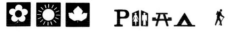

44 | Snake River Birds of Prey Area

Description: The SRBOPA was established in 1980 to protect a unique desert environment that supports North America's densest concentration of nesting birds of prey. More than 700 pairs of raptors nest each spring along 81 miles of the Snake River Canyon. Of the 14 documented breeding species here, the prairie falcon, golden eagle, red-tailed hawk, and northern harrier are the most common. Great horned owl, Swainson's and ferruginous hawks, American kestrel, turkey vulture, and the common barn, western screech, northern saw-whet, burrowing, short-eared, and long-eared owls also dwell here. Eight other raptor species migrate through the area. Basalt cliffs that tower up to 600 feet above the river, providing countless crevices for nesting, and dense Townsend's ground squirrel populations make this area great for raptors. Riparian bottomlands provide nesting habitat for songbirds (watch for Lazuli bunting and Say's phoebe) and cover for rodents and reptiles. The surrounding lands, dominated by several desert shrub species with a grass understory, foster dense ground squirrel and jackrabbit populations. The SRBOPA has one of the nation's highest densities of badgers and is one of the few places in Idaho to see black-throated sparrows. In all, more than 250 species of mammals, birds, amphibians, reptiles, and fish are found in the area including several lizard and snake species found nowhere else in the state: the Mojave, black-collared, leopard, side-blotched, and western whiptail lizards and the night, western ground, and longnose snakes. Mid-March to the end of June is the best time to see raptors; by July, scarce prey and high temperatures cause most raptors to leave the area.

Directions: Visitors may tour this area on their own by passenger vehicle or boat, or take a commercial guided tour or float trip. From Boise on Interstate 84, take Idaho 69 (exit 44) south for eight miles to Kuna. Continue south on Swan Falls Road following signs to the SRBOPA. The area begins five miles south of Kuna; at mile 15 there is a signed parking lot with a short trail for canyon rim viewing. Swan Falls Dam is 18 miles from Kuna. A dirt road along the river bottom leads downstream from the dam for several miles and also from the Grandview area. The BLM office in Boise at 3948 Development Ave. has maps and a 52-page color brochure on the area. Boaters are requested to register at the office or at launch site boxes.

Ownership: BLM (334-1582)
Size: 482,640 acres **Closest Town:** Kuna, Grandview

Up to 200 pairs of prairie falcons nest on cliff ledges in the Snake River Birds of Prey Area. These falcons can reach speeds in excess of 100 mph when diving on prey or defending their nests. D. BOIKE

45 Ted Trueblood Wildlife Area

Description: This area on the north side of the Snake River has three shallow ponds that attract thousands of migratory waterfowl, wading birds, and shorebirds. Islands and heavily vegetated areas adjacent to the ponds provide attractive nesting habitat for Canada geese and several species of ducks. Common nesting shorebirds and wading birds include American avocets, black-necked stilts, common snipe, and killdeer. Yellow-headed and red-winged blackbirds are abundant in the marshy areas while various warblers, sparrows, and other passerines use the dense riparian thickets. Burrowing owls, long-eared owls, and northern harriers are the most common nesting raptors. In the winter, watch for ring-necked pheasants and California quail, plus tundra swan, and other waterfowl. Golden eagles, red-tailed, Swainson's and Cooper's hawks, and great horned owls frequently search for prey over the area. It is a popular hunting spot.

Directions: Ted Trueblood Wildlife Area is located 0.5 mile north of Grandview along the west side of Idaho 67, on the north side of the Snake River. From Mountain Home, take Idaho 67 south for about 22 miles. There are three parking areas around the site perimeter: one along Idaho 67 and the other two along the river access road that forms the northern boundary. Visitors should call the C.J. Strike WMA for information.

Ownership: BLM, managed by IDFG (845-2324)
Size: 320 acres   **Closest Town:** Grandview

46 C. J. Strike Wildlife Management Area

Description: This reservoir fed by the Snake and Bruneau Rivers is bordered by marshes, ponds, and wildlife food plots. Ducks, geese, wading birds, and shorebirds number in the thousands during migration periods. Many raptor species nest in the area. White-tailed and mule deer are commonly seen at dawn and dusk. Viewers can drive on dirt roads that lead to the shoreline, walk or bike on closed roads, or boat the reservoir.

Directions: From Mountain Home, head south on Idaho 51 for 21 miles to Bruneau. Turn right onto Idaho 78 for about four miles to the WMA headquarters entrance sign and follow a dirt road for 0.5 mile to the end. Pick up a map at the WMA headquarters showing the various boat ramps, campgrounds, and access points to the reservoir shoreline. This site is a wildlife production, hunting, and fishing area, so there are seasonal access restrictions to portions of the area. Marsh areas are closed during the February 1 to May 31 nesting season.

Ownership: Idaho Power Co., managed by IDFG (845-2324)
Size: 12,500 acres   **Closest Town:** Bruneau

47 Bruneau Dunes State Park

Description: This state park boasts the tallest single structured sand dune in North America, at 470 feet tall. The dunes are intermixed with sagebrush desert and grassland flats. There are trails around two shallow, marshy lakes lined with riparian shrubs and trees where birds are abundant during migration. Most duck species that travel through Idaho can be seen here and many stay the winter. Water bird species include Canada geese, tundra swans, dabbling and diving ducks, and great blue herons. Bald eagles are present in the winter. A five-mile hiking trail starts at the visitor center and circles the park, and an excellent interpretive center has wildlife displays. Additional ponds north of the park are good for viewing shorebirds during migration, including the American avocet, black-necked stilt, killdeer, long-billed curlew, red-necked and Wilson's phalaropes, and western and least sandpipers. The park also abounds with mammals, amphibians, and reptiles, although often only their tracks are seen. Watch for coyote, black-tailed jackrabbit, Ord's kangaroo rat, short-horned and western whiptail lizards, and the gopher snake in the early morning or evening hours. Eagle Cove Environmental Education Center is an on-site, indoor program available to schools and other groups interested in learning more about the area.

Directions: From Mountain Home, travel south for 15 miles on Idaho 51 to the Snake River. Cross the bridge and turn left onto Idaho 78 for two miles to a signed road leading to the park entrance, which is one mile further. Pick up a guide to the park at the visitor center.

Ownership: SP (366-7919)
Size: 4,800 acres **Closest Town:** Bruneau

Yellow-headed blackbirds stake out their territories in early spring by singing from prominent perches and fluffing out their bright yellow feathers. DENNIS AND MARIA HENRY

SITE	SITE NAME
48	**Thorn Creek Reservoir**
49	**Hagerman Wildlife Management Area**
50	**Thousand Springs Preserve**
51	**Niagara Springs Area**
52	**Rock Creek Canyon/Shoshone Basin**
53	**Milner Reservoir**
54	**Minidoka National Wildlife Refuge**
55	**City of Rocks National Reserve**

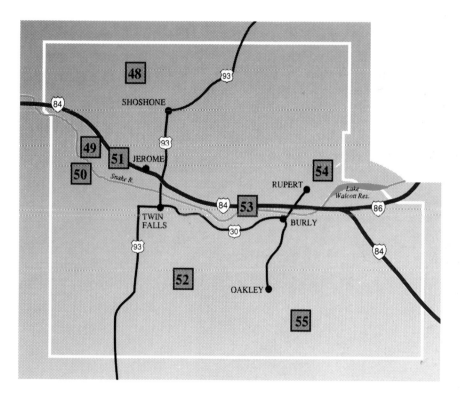

| 48 | Thorn Creek Reservoir |

Description: Thorn Creek Reservoir is surrounded by desert vegetation types distinctive of the Great Basin including sagebrush and grasslands among rolling hills. In the spring and fall waterfowl and shorebirds concentrate at the reservoir where Canada geese nest atop man-made platforms. In the early morning and late afternoon hours elk and mule deer drink at the shoreline and feed on shrubs and new grass growth. Sage grouse may be seen strutting during early mornings in the spring and brooding young throughout the summer. Mountain bluebirds migrate through the area in the fall. A sign at the dam explains the current BLM pilot riparian study. The reservoir provides excellent fishing opportunities through mid-summer and is open to non-motorized craft. To further explore this portion of Idaho, contact the Shoshone BLM office for drive routes through the Bennett Hills area and for information on viewing sage grouse at lek sites (traditional display grounds) in the area. Lek sites are not always reliable from year to year and you must be willing to be there at dawn in early spring.

Directions: From Gooding, head north for 17 miles on Idaho 46 and turn right at the sign for Thorn Creek Reservoir. In three miles you will come to a "T" intersection. Bearing right will lead you to the dam in 0.5 mile while going left 0.5 mile will take you to the upper west shoreline. The road is not winter-maintained and is poor when wet. Walk, mountain bike, or canoe around the reservoir, where there are primitive boat ramps.

Ownership: BLM (886-2206)
Size: 128 acres **Closest Town:** Gooding

The Columbian ground squirrel inhabits open forests and mountain meadows. Like all ground squirrels, they use cheek pouches to carry seeds and fruits to their underground burrows. TIM CHRISTIE

| 49 | Hagerman Wildlife Management Area |

Description: U.S. 30 in the Hagerman Valley is known as the "Thousand Springs Scenic Route." Here the disappearing Lost River of east-central Idaho returns to spew forth a series of white waterfalls down black canyon cliffs. The Snake River Plain aquifer is a massive underground system of lakes and "lost rivers" more than 150 miles long. The clear spring water, at 58 degrees, is just the right temperature for trout farming. In fact, 90% of the nation's commercial trout are raised in the valley. The WMA's marshy ponds surrounded by emergent aquatic vegetation provide great habitat for waterfowl and wading birds. Common species include Canada goose, mallard, ring-necked duck, lesser scaup, American widgeon, pied-billed and western grebe, American avocet, and black-crowned night heron. Raptors seen in the area include the northern harrier, bald eagle, great horned owl, and barn owl. Among the less common species are the common loon, black-necked stilt, tundra swan, and Forster's tern. The adjacent uplands are managed for game birds. Fishing and hunting are popular here. Nearby is the Hagerman National Fish Hatchery, where rainbow and steelhead trout can be viewed in indoor and outdoor rearing units. There are interpretive displays, a brochure, and a self-guided tour.

Directions: From Hagerman, follow U.S. 30 for 3.3 miles south and turn left onto a gravel road just past the rest area. Viewing is mainly by driving the roads but you can also walk the wide, dry dikes between July 15 and October 31. The state fish hatchery is adjacent to the WMA. To reach the national hatchery, go three miles south of Hagerman on U.S. 30 and turn left onto the Hagerman-Wendell Road. Turn right in one mile and go another 1.5 miles to the hatchery.

Ownership: IDFG (324-4359), USFWS (837-4896)
Size: 877 acres **Closest Town:** Hagerman

The common snipe is a secretive bird of wet grassy areas, bogs, and marshes. It has a remarkable courtship flight that includes a power dive in which the vibrating outer tail feathers produce a series of rapid, hollow notes. JAN L. WASSINK

50 | Thousand Springs Preserve

Description: This Nature Conservancy preserve includes over two miles of spring creeks and three miles of Snake River frontage, with canyon walls up to 400 feet tall and riparian bottomland. On the preserve's northern end, a spectacular display of springs burst forth from the fissured, vertical walls of lava rock. This represents the last unaltered canyon wall spring in a system that once stretched for tens of miles along the Snake River. At the southern end is the 400-foot tall Sand Springs Falls and two pristine spring creeks, home to the largest known population of the rare Shoshone sculpin, a bottom dwelling fish dependent on the spring water. Golden eagles and prairie falcons nest on the cliffs and hundreds of herons nest and feed along the dense cover that lines the many marshes and sloughs. Waterfowl are numerous in winter and migration periods. A good way to see the site and other portions of the Snake River in the Hagerman Valley is by canoe. For Thousand Springs, there is an upstream boat launch at the south end of West Point Road and a take-out site at the Idaho Power picnic area. Stay in your craft when passing active heron rookeries.

Directions: From the Hagerman National Fish Hatchery (see site No. 49), continue southeast for three miles to the Idaho Power Plant Picnic Area on the Snake River. Cross the bridge to the parking lot. Tours are arranged by appointment only (536-2242).

Ownership: The Nature Conservancy (536-2242)
Size: 425 acres **Closest Town:** Wendell

51 | Niagara Springs Area

Description: The Niagara Springs area includes an IDFG WMA, which provides excellent views of whitewater rapids on the Snake River, steep canyon cliffs, and a well developed riparian ecosystem. Spring and summer wildlife include the canyon wren, common yellowthroat, yellow warbler, northern oriole, yellow-breasted chat, Say's phoebe, wood duck, white pelican, great blue heron, double-crested cormorant, Canada goose, Caspian tern, golden eagle, and mule deer. This is a major waterfowl wintering site with over 5,000 birds often observed. To the east of the WMA is Niagara Springs, a National Natural Landmark where springs bubble and froth as they exit the Snake River aquifer and roar down the canyon wall.

Directions: From Interstate 84, take the Wendell exit 157 and go south for seven miles. The entrance road is signed. Park and walk through the WMA. Niagara Springs and Pugmire Park are on the WMA's eastern side. Facilities are at the park.

Ownership: IDFG (324-4359), SP
Size: 1,056 acres **Closest Town:** Wendell

52 | Rock Creek Canyon/Shoshone Basin

Description: Rock Creek Canyon offers easy viewing of mule deer on their winter range. The paved road travels through a steep rock-walled canyon along the riparian lined Rock Creek, where you might see mallards and teal. Deer are highly visible from December through February. Although winter wildlife viewing is limited, the Magic Mountain area has several cross-country ski trails. Occasionally a porcupine, long-tailed weasel, or Steller's or gray jay may be seen. May to October visitors may view wildlife at two other sites that lead to the Shoshone Basin. On the way to these sites, look along Rock Creek for the yellow warbler, American goldfinch, brown-headed cowbird, belted kingfisher, and northern flicker. The first site, near Electric Spring, is a 25-acre marsh fenced to protect nesting waterfowl and wading birds. Look for the USFS interpretive signs and for geese and ducks on the nesting islands. Heading west from the marsh into the sagebrush flatlands of the Shoshone Basin, pronghorn and sage grouse are often spotted. The forest has numerous campgrounds and hiking trails.

Directions: From Twin Falls, take U.S. 30 for nine miles east to Hansen. Turn south onto Rock Creek Road (there is a sign for the Magic Mountain Ski Area) for 10 miles to the mouth of Rock Creek Canyon. In winter, view deer for the next five miles. From May to October, continue on this road (which enters the Cassia Division of the Sawtooth National Forest) for 11 miles to the Magic Mountain Ski Area. Turn right onto Forest Road 500 for about 7.5 miles for marsh viewing. Continue west on Road 500. In four miles you will leave the National Forest, where the road turns to County Road G3. In another 12 miles you will be at the junction of U.S. 93 near Hollister. Twin Falls is 25 miles north. Contact the USFS in Twin Falls for information on ski trails. Maps are available in Twin Falls and Burley.

Ownership: BLM (678-5514), USFS (737-3200), PVT
Size: Five miles (winter), 40 miles **Closest Town:** Hansen

Contrary to popular belief, porcupines cannot throw their quills. When approached too closely they will slap their tails on the attacker, driving in dozens of barbed quills. MICHAEL S. SAMPLE

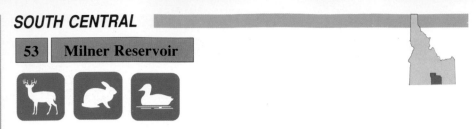

53   Milner Reservoir

Description: Milner Reservoir is surrounded by sagebrush-grassland and lined with basaltic cliffs along the northern shoreline. Viewing of water-associated wildlife is year-round. See American white pelicans in spring and summer and tundra swans in winter. Canada geese nest in area juniper trees. Migration periods are best for seeing the mallard, gadwall, teal, ruddy duck, American widgeon, and Canada goose. Ringed-necked pheasant, black-tailed jackrabbit, Nutall's cottontail, and Townsend's ground squirrel are also very common. Mule deer winter in the area.

Directions: From Burley, take U.S. 30 west for seven miles. At the Milner Historic/ Recreation Area sign turn north for one mile, then travel west for approximately three miles on a gravel road to the parking area. From the parking area, a sign directs visitors to the boat ramp a short distance to the east, where there is another parking lot and a picnic area. You can walk the shoreline, boat on the lake, or travel three miles of dirt road in an easterly direction.

Ownership: BLM (678-5514)
Size: 2,000 acres **Closest Town:** Burley

54   Minidoka National Wildlife Refuge

Description: This NWR extends for 25 miles along the Snake River from Minidoka Dam and includes all of Lake Walcott. The reservoir has several islands and marshy areas. The topography is primarily low, rolling hills and lava rock ledges up to 30 feet high along the shore. Migratory waterfowl are the most prominent wildlife on the refuge, an important stopover point in the Pacific Flyway. Flocks of over 500 tundra swans, along with 27 other waterfowl species have been recorded. The refuge has Idaho's only nesting white pelicans, with over 40 pairs. Many songbirds and raptors nest or migrate through the refuge, which also hosts a wide variety of mammals including mule deer, beaver, muskrat, coyote, and Nutall's cottontail. Pronghorn are present in small numbers.

Directions: From Rupert, drive six miles northeast on Idaho 24. Proceed through Acequia, then turn east for six miles on County Road 400 N. to the refuge headquarters located on the north side of the Minidoka Dam. The restrooms, picnic area, and campground are in Walcott Park, which is administered by the USBR. There are three south shore access points for vehicles. Sedans are not recommended for the northern access roads.

Ownership: USFWS (436-3589), USBR (436-6117)
Size: 20,721 acres **Closest Town:** Rupert

| 55 | City of Rocks National Reserve |

Description: The City of Rocks gets its name from the grotesque, eroded granite formations of sheer cliffs and pinnacles towering up as much as 60 stories above the valley floor. Many settlers along the California Trail wrote their names in axle grease on these rocks. The dominant vegetation is an extensive stand of pinyon pine, juniper, and mountain mahogany, with occasional aspen and whitebark pine. Songbirds are the predominant attraction here, featuring the pinyon and scrub jay, green-tailed towhee, Virginia's warbler, mountain bluebird, Clark's nutcracker, and Townsend's solitaire. Also found here are turkey, prairie falcon, burrowing owl, poor-will, Say's phoebe, white-throated swift, black-chinned hummingbird, house, canyon, and rock wrens, common bushtit, gray flycatcher, plain titmouse, and red-naped sapsucker. This site is world famous for its challenging rock climbing. Reserve roads are typically closed due to snow from December 1 through March.

Directions: *From Interstate 84 east of Burley, exit onto Idaho 77 south and go about 20 miles to Conner. Turn right at the main intersection and go 16 miles to Almo. Go one mile past Almo, turn right, and drive four miles. Or, take Idaho 27 to Oakley, turn east for one mile, then turn south on Birch Creek Road for 14 miles (graveled roads). Turn east up Emery Canyon Road for two miles to the City of Rocks. Camping is allowed but specific sites are not designated.*

Ownership: NPS (733-8398), SP, PVT
Size: 14,400 acres **Closest Town:** Almo, Oakley

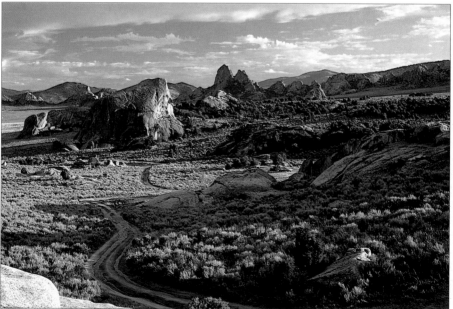

The scenic pinyon pine-juniper woodland at City of Rocks provides opportunities for viewing several songbirds uncommon to Idaho, such as the pinyon and scrub jay, and Virginia's warbler. BOB MOSELEY

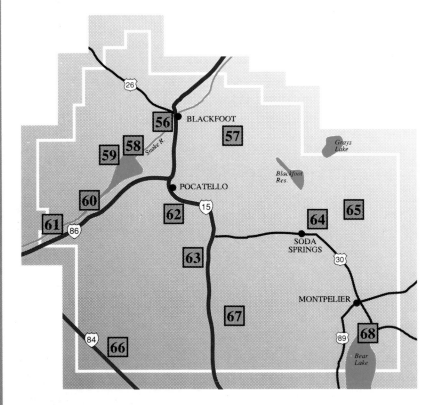

56 | Snake River at Blackfoot

Description: Below Blackfoot the Snake River is lined with dense cottonwood stands and flows through agricultural and sagebrush habitats. Prominent spring to fall wildlife include bald eagles, great blue herons, osprey, mergansers, Canada geese, mule and white-tailed deer, beaver, river otter, and a few waterfowl and shorebird species. There is a heron rookery along this route, so observers should remain in their craft when near the nests in the spring and early summer months. During the winter, the river freezes and is not floatable.

Directions: From Interstate 15 at Blackfoot take Exit 93. Turn right towards town and make a right turn at the first light onto Parkway Drive. Proceed about 0.25 mile and turn right onto W. Bridge Street for about 0.5 mile to the river. The boat launch is just upstream from this road. From here you can boat or raft down the river for about 10.5 miles. Take out on the downstream side of Tilden Bridge, on the right shore. To return to Interstate 15 head east on Ferry Butte Road from the bridge. You can also take out three miles upstream of the bridge at the Riverton Sportsman Access Point. From the Riverton landing go south for about two miles to where the road joins with Ferry Butte Road.

Ownership: BLM (529-1020), IDL, PVT
Size: 10.5 miles of river   **Closest Town:** Blackfoot

Wetlands, such as marshes, ponds, lakes, and streams are required habitat for many of Idaho's wildlife species, such as this great blue heron. Loss of wetlands poses a signifi-cant threat to many wildlife populations. RALEIGH MEADE

Description: This portion of the Blackfoot River canyon is a smaller version of the Snake River Birds of Prey Area (site #44). Steep canyon cliffs and cottonwood and aspen trees support nesting golden eagles, prairie falcons, red-tailed hawks, great horned owls, and other raptors. Watch for typical sagebrush inhabitants including sage sparrows and sage and sharp-tailed grouse along the canyon rim. Songbirds here are abundant and diverse—over 100 bird species have been observed during the springtime. Look for gray catbirds, black-throated gray warblers, blue-gray gnatcatchers, plain titmice, pinyon jays, and green-tailed towhees. Elk, mule deer, and moose are occasionally seen along the river in the summer. Viewing for the first 10 miles is from the rim of the canyon. In the next 17 miles there are six spots to explore along the river. Rafting the river is another way to see the wildlife. Plan to visit the area between May and November as the road conditions are poor in winter and early spring.

Directions: From Blackfoot, take U.S. 91 north about seven miles and turn right onto Wolverine Road. Head east for about 10 miles to Wolverine Creek, where the route starts. Turn right and cross the creek onto Cedar Creek Road. Follow this road for the next 10 miles as it parallels the river canyon's rim. It will turn into Blackfoot River Road after about four miles. Park in turnouts along the road and walk approximately 200 yards to the rim for viewing. There are a few road spurs where you can drive in closer. There are six locations that access the canyon bottom. Continuing on the same road, you will come to Trail Creek Road in about three miles; turn right onto it and cross the river. In six more miles you will pass Morgans Bridge (Paradise Road), on your left. The road then turns into Lincoln Creek Road and over the next six miles you will pass Graves Creek, Cutthroat Trout, and Sagehen campgrounds, which are all along the river's edge. Two miles farther at Negro Creek turn left and go 0.5 mile to the river. The road continues to Blackfoot Reservoir in about seven miles and is a good connecting route to Grays Lake National Wildlife Refuge. A popular rafting run is from Morgans to Trail Creek Bridges. Facilities are at the campgrounds.

Ownership: BLM (529-1020), PVT
Size: 30 miles of river **Closest Town:** Blackfoot

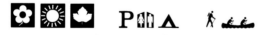

Many birds migrate at night. They use the setting sun or the stars to set a course, and some species use the earth's magnetic field to guide them on their journeys.

| 58 | Springfield Bottoms—American Falls Reservoir |

Description: From mid-July through September these extensive spring-fed mudflats may be the best place in the state to view thousands of shorebirds. Over 8,000 birds representing more than 30 species were counted in the fall of 1988. Abundant species include killdeer, American avocet, western and Baird's sandpiper, and long-billed dowitcher. Some of the uncommon to rare species include black-bellied, lesser golden, and semipalmated plover, greater yellowleg, willet, whimbrel, ruddy turnstone, red knot, stilt sandpiper, and short-billed dowitcher. Prairie and peregrine falcons are often present when prey numbers are high. Abundant waterfowl replace the shorebirds in the spring. The resident northern saw-whet, western screech, and great horned owls can be found on the juniper and cottonwood-covered McTucker Island.

Directions: From Springfield, travel east on Idaho 39 for approximately 2.5 miles to Stecklein Road. Turn left and go 2.8 miles to River Road (gravel). Turn right and drive 0.9 mile where the road bends sharply to the right and there is a left fork. Go left across the cattle guard. Cross two more cattle guards and turn right onto a two-rut dirt road as far as you can drive (about one mile). If the road is muddy, it is best to stop and walk to the mudflats. Taking the left fork at the last intersection leads to McTucker Island. The river can be waded except during spring high water.

Ownership: USBR (226-2217)
Size: Two miles of river **Closest Town:** Springfield

| 59 | Springfield Ponds |

Description: These spring-fed ponds surrounded by shrubby riparian vegetation and marshland are great for bird watching. Most waterfowl species can be viewed here during migration. Surf and white-winged scoters, harlequin and wood ducks, oldsquaw, greater scaup, and trumpeter swan are among the uncommon waterfowl, while common loon, Virginia rail, sora, common yellowthroat, yellow and yellow-rumped warblers, and black-headed grosbeak are often seen. You may also spot Pacific loon, varied thrush, and Harris' sparrow.

Directions: From Blackfoot, travel southwest on Idaho 39 for approximately 20 miles to Springfield. The ponds are about 0.5 mile west of town. Use road pullouts for viewing. There is a city park with picnic facilities in Springfield.

Ownership: City of Springfield
Size: 25 acres **Closest Town:** Springfield

| 60 | **American Falls Dam and Vicinity** |

Description: The entire American Falls Dam area is good for viewing bald eagles and water birds, including common loons, in the winter. The bald eagles communally roost in juniper stands in the cemetery; leave that site well before dusk to avoid disturbing incoming birds. Peregrine falcons, parasitic jaegers, and Sabine's and Thayer's gulls are also winter visitors, although rare. During migration over 30 species of shorebirds have been recorded on the mudflats near the silo. In the summer white pelicans, double-crested cormorants, white-faced ibis, terns, and gulls are common.

Directions: This site includes three separate viewing areas along the south bank of the Snake River both below and above American Falls Dam. From Interstate 86 west of American Falls take the American Falls-Rockland Exit (Idaho 37) and turn left under the freeway. To visit the Pipeline Recreation Area turn left at the first intersection and head southwest for on the frontage road 1.5 miles, paralleling Interstate 86. Turn right at the sign onto a gravel road for 0.5 mile to the parking lot. During snowy conditions access is limited to 4-wheel drive vehicles. Return to Idaho 37, go about 1.5 miles past the freeway interchange and turn left onto Teton Ave. for about 0.5 mile. Then turn left onto McKinley St. and turn right into the Falls View Cemetery, which overlooks the Snake River. To visit the site above the dam return to Idaho 37 and continue into American Falls. Turn left at the first stoplight onto Idaho 39 in downtown, go 0.5 mile, and make a right turn onto a gravel road before the dam. It is a few hundred yards to the reservoir shoreline, where you can see an old silo on the water's edge. Park and walk along the shoreline.

Ownership: USBR (226-2217), BLM (678-5514), City of American Falls
Size: Three miles of river **Closest Town:** American Falls

| 61 | **Massacre Rocks State Park** |

Description: This sagebrush desert area, dotted with cindercones, sits adjacent to the Snake River just five miles upstream of the Minidoka NWR and is used by many of the same wildlife species. Bird watching in spring and fall is excellent. Mammals are less commonly seen, although you may spot the tracks or other signs of deer, elk, coyote, badger, bobcat, white-tailed jackrabbit, striped skunk, or porcupine. Commonly seen snake and lizard species include the western terrestrial and common garter snakes, sagebrush lizard, and western skink.

Directions: From American Falls, go nine miles west on Interstate 86 to Exit 28 and follow the signs. Visit the park headquarters for trail maps and bird checklists.

Ownership: SP (548-2672)
Size: 900 acres **Closest Town:** American Falls

62 Cherry Springs Nature Area

Description: This site includes three self-guided nature trails with over 50 interpretive signs, two learning centers, and an amphitheater. The trail guide provides wildlife, botanical, and geological information, and asks questions that encourage active observation and interpretation. The trails wind through dense riparian vegetation bordered by mountainous sagebrush-grassland and juniper habitats. Of the over 100 documented bird species, common nesting species include hermit and Swainson's thrush, dusky, gray, and Hammond's flycatchers, Virginia and black-throated gray warblers, green-tailed and rufous-sided towhees, and poor-will. During fall and winter look for the American robin, ruby-crowned kinglet, Bohemian and cedar waxwings, solitary and warbling vireos, dark-eyed junco, and golden eagle. Thirty mammal and several reptile and amphibian species inhabit the area, although tracks or other signs may be the best way to "see" them. Common mammals include the least chipmunk, red squirrel, white-tailed jackrabbit, raccoon, western spotted skunk, coyote, and red fox. The sagebrush lizard, western skink, and gopher, garter, and western rattlesnakes are often seen. To explore beyond the Nature Area drive up Mink Creek, East Fork Mink Creek, or South Fork Mink Creek where many trailheads are located. The Mink Creek drainage is a very popular winter sports area. Cherry Springs nordic trails are excellent for beginners, while the trails of the East, West, and main Forks of Mink Creek range from "easiest" to "most difficult." Nordic trails other than at Cherry Springs require "Park n' Ski" permits.

Directions: From Pocatello, head southeast on Bannock Hwy. for four miles and turn south onto Forest Road 231, which follows Mink Creek. Go four miles to the signs for the Nature Area. Birding is good along Mink Creek and its major tributaries—the East, West, and South Forks, plus Kinney Creek. The Mink Creek drainage ends at Crystal Summit, approximately 4.5 miles past the Nature Area. Pick up a trail guide and wildlife checklists at the information shelter in the Nature Area. USFS maps are available at the Pocatello Ranger District office.

Ownership: USFS (236-7500)
Size: 180 acres **Closest Town:** Pocatello

This red squirrel is eating a mushroom, one of its favorite foods. It will also eat nuts, buds, flowers, lichens, bark, and conifer seeds.
TOM AND PAT LEESON

63 | Hawkins Reservoir

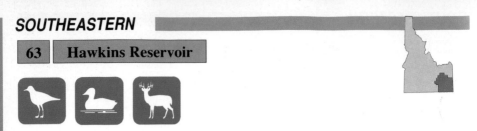

Description: This scenic and easily accessible mountain reservoir offers open water and mudflats, attracting both waterfowl and shorebirds. It is set in juniper and mountain shrub habitat with aspen, Rocky Mountain maple, and Douglas-fir stands as backdrop. Late spring through early fall is the best time to see plovers, sandpipers, ducks, and geese. Shorebirds are best viewed by walking across the dam to the south shoreline. You can also boat the reservoir or drive for one mile along the north shoreline on the main road. Sharp-tailed grouse nest in the area and mule deer, coyote, porcupine, and red fox are area residents. While moose and elk are rare, they are most likely seen in the spring and summer.

Directions: *From Virginia on Interstate 15, take Exit 36 and head west 8.5 miles on Virginia Road to the reservoir.*

Ownership: BLM (766-4766)
Size: 120 acres **Closest Town:** Virginia

64 | Formation Springs Preserve

Description: This Preserve protects crystal clear pools and a unique wetland complex at the base of the scenic Aspen Mountains. The cold springs that feed the terraced pools and creek system deposit high concentrations of travertine (calcium carbonate), which gives the site its unique geology. The spring water has been determined to be 13,000 years old. Perhaps the most impressive physical feature is Formation Cave, which is almost 20 feet tall at the entrance and 1,000 feet long. The ponds attract numerous wintering waterfowl and support a healthy trout population. On the preserve look for elk, mule deer, raptors, and numerous songbirds.

Directions: *From Soda Springs, take Idaho 34 north for two miles, turn right onto Forest Road 124 (at the stoplight), and go one mile to the Preserve, which will be on your left.*

Ownership: The Nature Conservancy (726-3007), BLM
Size: 160 acres **Closest Town:** Soda Springs

 Wildlife can be identified by the signs they leave such as tracks in mud and snow, beaver gnawings, owl pellet castings, and bear scratchings.

Idaho's abundant elk herds attract a large number of big game enthusiasts from throughout the country. Elk are also one of Idaho's most popular wildlife-viewing species. Bull elk can weigh up to 1,000 pounds. TIM CHRISTIE

65 | Diamond Valley/Elk Valley Marsh

Description: This tour route follows scenic mountain creeks bordered by aspen trees and surrounded by coniferous forest—a good example of eastern Idaho's national forest habitats and wildlife. Beaver dams, which create waterfowl ponds, are frequently spotted and the state's highest densities of moose are found in this area. Willow flats often hide foraging moose in the early morning and evening hours. Also watch for elk and mule deer. Significant numbers of spawning cutthroat trout can be seen from late May to early June in Diamond Creek, where the Forest Service has installed hunderd bank support structures, fencing and pool-creation structures to improve and protect important fish habitat. Elk Valley Marsh is a remote, high-altitude, 200-acre mountain marsh bordered by sagebrush-grassland, conifer forest, and scattered aspen stands. It is used as a nesting and molting area for Canada geese and dabbling ducks. Sandhill cranes, moose, mule deer, and sometimes elk can also be seen here. A trailhead starting at Road 147 offers a five-mile loop trail trough the Gannet Hills.

Directions: The tour can be driven as a loop with a short side trip or one-way toward other sites in the state. From Soda Springs, follow Idaho 34 north for 11 miles and take the Blackfoot River Road heading east. In seven miles the road will enter the National Forest. The Diamond Creek drainage begins in aproximately 10 miles. You will be on Forest Road 102 and will paralell the creek for aproximately 12 miles to the head of Georgetown Canyon and Wells Canyon (Road 146). Staying on Road 102 will take you to Georgetown in approximately 11 miles (head north on U.S. 30 to return to Soda Springs). Taking the Wells Canyon route will lead you to Montpelier. Follow Road 146, which will end in four miles, and turn south onto Road 111 (Crow Creek Road). In another five miles you will come to the junction of road 147. This road leads to Elk Valley Marsh in six miles. Park along the road and scan the marsh to the east. Returning to Road 111, it is another 10 miles south to U.S. 89 and an additional 6 miles west to Montpelier. On this route you will pass five signed trailheads. Mill Canyon and Diamond Creek Campgrounds have restrooms. There are three more campgrounds near Montpelier Reservoir. This tour is almost entirely on gravel roads and is not recommended for sedans or station wagons during wet weather or in the winter. A Caribou National Forest map is recommended and can be purchased in Soda Springs.

Ownership: USFS (547-4356)
Size: 60-to 80-mile loop tour, 20-to 40-mile one-way
Closest Town: Soda Springs, Montpelier

In newly burned forests, birds such as pine siskins, Clark's nutcrackers, and red crossbills move in to feed on millions of seeds released by lodgepole pine cones during the searing heat.

66 | Juniper Rest Area

Description: This rest area offers easy access in a frequently traveled area for viewing several juniper-associated wildlife species. It is set in the old lake bed of Lake Bonneville. You can view from the parking lot or venture off on foot paths leading west or south of the rest stop. In winter look for plain titmice, pinyon and scrub jays, and rough-legged hawks. In the spring watch for the vesper sparrow, mountain bluebird, ferruginous hawk, red-tailed hawk, and northern harrier.

Directions: *Juniper Rest Area is on Interstate 84, six miles south of Juniper and five miles north of the Utah border.*

Ownership: BLM (766-4766), ITD
Size: 40 acres **Closest Town:** Juniper

67 | Oxford Slough/Twin Lakes/Swan Lake

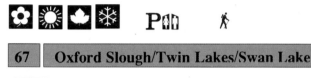

Description: Oxford Slough is a freshwater cattail-bulrush marsh with seasonally flooded alkali flats that support colonial nesters including the white-faced ibis, Forster's and black tern, eared grebe, snowy egret, and black-crowned night heron. It is a waterfowl production area, especially for redheads. Trumpeter swans and whooping cranes have occasionally been observed. Surrounded by cottonwoods and dry hills, Twin Lakes is a staging area for the common loon. Swan Lake, a shallow cattail-ringed lake, is an important spring stopover point for tundra swans and thousands of ducks. Plain titmice, black-throated gray warblers, blue-gray gnatcatchers, and gray flycatchers can be observed in the surrounding juniper foothills. Rough-legged hawks and northern harriers can be seen throughout the valley in the winter.

Directions: *From Red Rock Pass, seven miles south of Downey on U.S. 91, turn south on paved County Highway D1 for seven miles to Oxford. View Oxford Slough from the road for three miles. Continue one more mile to a Sportsman's Access sign directing traffic to Twin Lakes, turn left on the gravel road, and travel 2.5 miles. The county park road follows the west, north, and east shoreline. From Twin Lakes, continue to the north and east on the gravel road until it rejoins U.S. 91 near Banida. Turn north on U.S. 91 to Red Rock Pass to complete the loop trip. You will pass Swan Lake just west of the highway about one mile south of the town of Swanlake. There is a pullout with a wildlife interpretive sign for the lake. USFS maps are available in Preston and the Swanlake General Store has maps of the local area.*

Ownership: BLM (766-4766), portions administered by USFWS, Franklin County, PVT
Size: 5,000 acres **Closest Town:** Clifton, Oxford

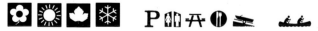

| 68 | **Bear Lake National Wildlife Refuge** |

Description: This NWR lies in the mountain-ringed Bear Valley at elevations ranging from 5,925 feet on the marsh to 6,800 feet on the rocky slopes of Merkley Mountain. It encompasses what is locally known as Dingle Swamp or Dingle Marsh and is comprised mainly of bulrush-cattail marsh, open water, and flooded meadows of sedges, rushes, and grasses. The dense bulrush stands provide nest material, breeding sites, and concealment for a diversity of migratory birds and small mammals. April and May are ideal months to view incoming waterfowl in breeding plumage. The Salt Meadow Wildlife Observation Route is a good area to observe a high diversity of birds including Canada geese, redheads, sandhill cranes, willets, and Wilson's phalaropes. The refuge harbors one of North America's largest nesting colonies of white-faced ibis. In the summer, western grebes, double-crested cormorants, gadwall, and Franklin's gulls become abundant. In September watch for 100-150 sandhill cranes feeding on refuge grain fields. Severe winter weather forces most waterfowl to migrate south by late November, although you may see some common and hooded mergansers, goldeneyes, and bald eagles if the water remains open. Mule deer winter by the hundreds along Merkley Mountain, and one or two elusive moose are present year-round. Other mammals include muskrats, striped skunks, and Nutall's cottontail. The refuge is open to winter cross-country skiing wherever hiking is permitted.

Directions: *From Montpelier, follow U.S. 89 west for three miles and turn left at the refuge directional sign onto Bear Lake County Airport Road. Drive five miles to the refuge entrance road. Pick up a brochure with a map and a bird checklist on-site or at the refuge headquarters in Montpelier. The restrooms, campground, and picnic area are at North Beach State Park on the southern boundary of the refuge. At times, large portions of the refuge are closed to protect wildlife during sensitive periods. Most interior roads are open to vehicle traffic from September 25 through January 15. Hiking is not permitted from March 1 through June 30 to protect nesting birds.*

Ownership: USFWS (847-1757)
Size: 18,000 acres **Closest Town:** Montpelier

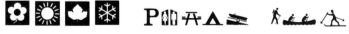

A cottontail suns itself on a winter day. Rabbits are an important food item in the diets of golden eagles, red-tailed hawks, great horned owls, red fox, bobcats, lynx, and coyotes.
TIM CHRISTIE

EASTERN

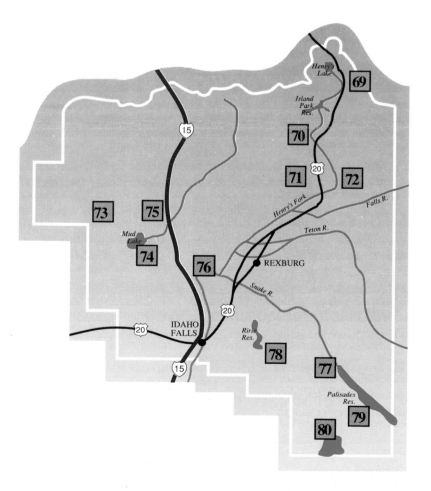

69 | Big Springs

Description: This is one of the country's largest springs, producing 120 million gallons of water a day. It supports enormous rainbow trout, which feed at the spring and are protected from fishing. View the fish from a 0.5-mile trail; also look for osprey, bald eagles, waterfowl, and an occasional moose, elk, white-tailed deer, muskrat, and black bear. A water trail here for canoes and rafts could provide good wildlife watching during the uncrowded morning and evening hours. Wildlife can be seen year-round, although winter access is restricted by snow. The surrounding forest contains Idaho's best population of great gray owls. Those entering or leaving the state via U.S. 20 should watch for Henry's Lake Flat just north of this site. It is a large, open grassland currently used for private livestock grazing. In the summer and fall look for pronghorn as well as sandhill cranes, which leave by mid-September for their Mexico wintering grounds.

Directions: Take U.S. 20 to Island Park Ranger Station, go six miles north to Macks Inn, turn right onto Forest Road 59, and drive four miles to Big Springs. This is a loop road that returns to U.S. 20. To see Henry's Lake Flat, continue north on U.S. 20 and view for seven miles to the intersection with Idaho 87. This is a drive-by site with no turnouts.

Ownership: USFS (558-7301)
Size: Five acres **Closest Town:** Island Park

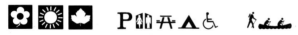

Great gray owls are large, rarely seen, forest dwelling owls that nest atop broken off trees and abandoned hawk nests. They hunt at night in forest openings and meadows for small mammals. TOM ULRICH

70 | Harriman State Park

Description: The Henry's Fork of the Snake River courses for six miles through this state park's 4,700 acres of lodgepole pine forests, meadows, and marshes, and past two small lakes. The Teton and Centennial (Continental Divide) mountain ranges are visible as well as the north escarpment of the 20-mile wide Island Park Caldera. Breeding pairs of trumpeter swans are the most conspicuous species here, and are present year-round. This area is also crucial for up to 5,000 wintering swans from Canada, Alaska, Wyoming, Montana, and Idaho. Spring through fall is the best time to see other wildlife, as winters can be harsh. Ten miles of hiking trails pass through meadow, forest, and lake habitats. In open areas and along water courses look for bald eagles, osprey, Canada geese, many duck species, long-billed curlews, and sandhill cranes (close to 50 crane pairs nest here). In the forested reaches watch for northern and black-backed woodpeckers, Williamson's sapsuckers, Steller's and gray jays, red crossbills, western tanagers, and the rarely seen great gray owl. Common mammals include elk, mule and white-tailed deer, moose, black bear, beaver, muskrat, river otter, yellowbelly marmot, badger, weasel, coyote, and fox. In winter, cross-country ski trails are groomed with facilities maintained year-round. Just south of the park is Swan Lake, an 80-acre marsh/pond where one pair of trumpeter swans nest annually. Four to six swans may be observed prior to nesting. Following nesting, from July through September, the adult pair and one to four cygnets are visible. Moose are also often seen from the turnout.

Directions: *From Ashton, travel 19 miles north on U.S. 20 and turn west 0.5 mile before the Osborne Bridge over the Henry's Fork. Two miles before the Park entrance you will pass Swan Lake with a small turnout for viewing. Maps and bird checklists are available at the park headquarters. Island Park Dam, eight miles north of the park, has good bird watching.*

Ownership: SP (558-7368) USFS (652-7442)
Size: 4,700 acres **Closest Town:** Island Park

Harriman State Park and adjacent rivers provide critical ice-free wintering habitat for up to 5,000 trumpeter swans. Several pairs are resident, raising three to six cygnets each year. HANK FISCHER

71 Sand Creek Wildlife Management Area

Description: The diverse geological features of this WMA include moving sand dunes, lava tubes, broken lava reefs, and cinder buttes. It is a high desert area with rolling hills of sagebrush-bunch grass rangeland interspersed with bitterbrush, chokecherry, shiny-leaf ceanothus, and other shrubs. The northern end contains lodgepole pine forest. The WMA and adjacent private lands are managed as winter range for 2,500 elk, 1,500 white-tailed deer, and 200 moose as well as for sage and sharp-tailed grouse. In the early spring watch for elk, moose, and white-tailed deer on migration, breeding waterfowl (including trumpeter swans), sandhill cranes, sage, sharp-tailed, blue, and ruffed grouse, golden eagles, osprey, prairie falcons, Swainson's and red-tailed hawks, turkey vultures, and a variety of songbirds. Occasionally you may see a bald eagle, great gray owl, common loon, muskrat, beaver, mink, yellowbelly marmot, white-tailed jackrabbit, red fox, porcupine, coyote, or pronghorn. In the summer, view the same species with their young, although opportunities are less frequent, while in the fall you can see most of the same species more frequently than in the summer. Many wildlife species can be viewed from either the roadway or from trails along the dikes. Sharp-tailed grouse will primarily be found around the sand dunes while sage grouse on leks can be viewed from several locations along Red Road, on the western edge of the WMA, or along the Sand Creek Road. Do not approach the leks on foot. The most popular spot is the northern end, especially the Sand Creek Ponds area, which has restrooms and primitive camping. The annual moisture is considerable, resulting in a wide variety and abundance of wildflowers in spring and summer. Fishing and hunting are popular. Non-motorized boats or other small craft are allowed on the ponds only after July 1 to protect nesting waterfowl. Trumpeter swan nesting areas are also closed to entry until July 1. Vehicle access is restricted in most areas in the winter to protect big game animals.

Directions: Take U.S. 20 to St. Anthony, exit onto Middle Street, and travel north through town. At the USFS office turn east and drive 1.5 miles. Turn left onto Sand Creek Road and go 16.5 miles to the Sand Creek Ponds. To reach the WMA headquarters, turn west at the USFS office onto North Parker Road, go four miles, turn right, and proceed 1.5 miles. A brochure and species checklists will be available on the WMA by the fall of 1990. Campsites are primitive.

Ownership: IDFG (624-7065), BLM, IDL
Size: 30,000 acres **Closest Town:** St. Anthony

Owls go unnoticed by their prey due to their silent flight. Very soft body feathers and the saw-toothed leading edge of their wing feathers allow for entirely silent flying.

Description: Idaho 47, from Ashton to the Harriman State Park area, is a highly scenic mountain drive designated by the USFS as a Scenic Byway. In the near future there will be 14 interpretive stops on scenery, forest practices, fish and wildlife resources, Upper and Lower Mesa Falls, and Harriman State Park. Look for brown, rainbow, brook, and cutthroat trout and whitefish from a fish observation platform where Idaho 47 crosses Warm River. The best viewing is in summer. Drop a few bread crumbs into the water to bring the fish to the surface. Osprey, bald eagles, and river otter frequent the area. Continuing north on Idaho 47, you will pass Upper and Lower Mesa Falls on the Henry's Fork, at 114 and 65 feet, respectively. You can easily see the lower falls from Grandview Campground; to reach the upper falls take Forest Road 295 west for 0.5 mile. Island Park Siding is a unique one-mile square area where the forest opens up into sagebrush flats. Pronghorn feed on the sagebrush and use the coniferous trees for cover. From the road scan these natural openings for pronghorn in the summer and sandhill cranes in the spring and fall. Also watch for moose (in shallow ponds and meadows), elk, and mule deer anywhere along this route. There are outstanding wet meadow wildflower displays from mid-May through June.

Directions: From Ashton on U.S. 20, take Idaho 47 east for eight miles to the Warm River fish observation platform. Continue on Idaho 47 for about 14 miles (you will pass the Mesa Falls overlook in five miles). Turn right onto Hatchery Butte Road and travel 4.5 miles. Turn left onto Forest Road 150 and go eight miles to Island Park Siding. Turn left onto Forest Road 291 and go three miles to return to U.S. 20. Island Park Reservoir is just two miles north.

Ownership: USFS (652-7442), ITD, managed by IDFG (522-7783)
Size: 37-mile drive **Closest Town:** Ashton

 **P**

Visitors to the Warm River Fish Observation site may see rainbow trout, shown here, plus brown, brook, and cutthroat trout. DAVID H. SAKAGUCHI

73   Birch Creek Valley

Description: Driving through this beautiful rangeland area between the Lemhi and Beaverhead mountain ranges provides a great opportunity to see pronghorn up close. They are often on or near the highway from spring to fall. Sage grouse are commonly seen in large groups from late summer through fall to the west of Idaho 28. Many dirt roads on BLM property traverse side canyons where the pronghorn and grouse can also be seen; Bartel Canyon, signed on the main highway, is a good one. A high-clearance vehicle is recommended. At the BLM's Birch Creek Campground, about six miles of primitive roads follow the creek with great rainbow trout fishing and wildlife viewing. The Birch Creek Springs Complex, at the north end of the valley, has an unusual alkaline spring-seep system that supports rare plants, waterfowl, songbirds, raptors, and pronghorn. *Primula alcalina* (a primrose that is a candidate for federal endangered status), an uncommon willow, and a lomatogonium occur in the area. Birdwatchers should look for willet, phalarope, snipe, long-billed curlew, savannah sparrow, sage thrasher, northern harrier, and red-tailed and ferruginous hawks. Kelsey's phlox and shooting stars are spectacular in spring and many other wildflowers contribute to a colorful show throughout spring and early summer. There are no established trails, but you can walk throughout the area. For further exploration of this region, Summit Creek in the Little Lost River Valley is another scenic, spring-fed stream. The 300-acre area has been fenced to exclude livestock since 1976, resulting in well developed riparian vegetation. In the spring and summer look for yellow warblers, willow flycatchers, American bittern, sage grouse, pronghorn, and rainbow trout. Walk up or downstream from the picnic area. Wildflowers are abundant and colorful. The rare plants mentioned above are also at this site.

Directions: From the intersection of Idaho 28 and Idaho 22, 50 miles north of Idaho Falls, head north on Idaho 28. The Bartel Canyon sign will be on your left in about four miles. You can drive in about six miles and will cross Birch Creek (probably a dry channel) in two miles. For about four miles up and downstream of the crossing large groups of sage grouse can be seen off the paralleling roads in the appropriate season. Returning to Idaho 28, continue north for 18 miles to Birch Creek Springs. You will pass the BLM Birch Creek Campground with restrooms in about 12 miles. Lone Pine, three miles further north, has gas, restrooms, and a cafe. Idaho 28 crosses Birch Creek about three miles after Lone Pine, where a dirt road on the right leads to an IDFG parking lot. The spring complex is on the north side of the lot. To access Summit Creek return to Idaho 22, head west for 25 miles, turn north at Howe onto a main paved highway, and go about 35 miles. There is parking at the picnic site, restrooms, and a BLM campground here. For both sites, confine motorized travel to existing roads to protect the rare plants.

Ownership: BLM (756-5400), USFS, IDFG, INEL, PVT
Size: 20 miles of river   **Closest Town:** Mud Lake, Leadore

     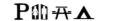

Idaho has 37 types of reptiles and amphibians. They can often be seen on rocks and logs basking in the sun or near water.

| 74 | Mud Lake Wildlife Management Area |

Description: This shallow lake, averaging five feet deep, is bordered by bulrush-cattail stands, saltgrass, and willows in an area surrounded by tall sagebrush desert, farmland, and pastures. This is an important spring water bird stopover along the Pacific Flyway. March to April are excellent times to view large numbers of snow and Canada geese, tundra and trumpeter swans, and many duck species. Other spring birds often seen include the American white pelican, Clark's and western grebe, common loon, black-crowned night heron, white-faced ibis, snowy egret, black tern, and double-crested cormorant. During the fall, migrating bird numbers are lower. Shorebird and songbird populations peak in May and many songbirds stay to nest. Large numbers of raptors nest in the area including red-tailed and Swainson's hawks, American kestrel, and northern harrier. In the winter look for bald eagles, peregrine falcons, and northern goshawks. Mule deer and pronghorn are present year-round. You can drive, launch small boats, or walk out on several points jutting into the lake. Fishing and hunting are popular here.

Directions: From Idaho Falls, drive north on Interstate 15 and turn west at Interchange 143 (Sage Junction). Follow Idaho 28-33 for about seven miles west to an intersection marked with a WMA direction sign. Proceed three miles north (where road turns to gravel), 0.5 mile west, one mile north, then 0.8 mile on a winding road to the lake. Maps and bird checklists are available at the WMA headquarters on the north side of the lake. There is a public viewing platform near nesting water birds.

Ownership: IDFG (522-7783)
Size: 8,853 acres **Closest Town:** Terreton

Pronghorn are found in shrub steppe habitat throughout the southern portion of the state. Birch Creek Valley is an excellent area for viewing as pronghorn are often close to the highway. CHRISTOPHER CAUBLE

83

75   Camas National Wildlife Refuge

Description: About half of this large refuge consists of lakes, ponds, and marshlands while the remainder is sagebrush-grass uplands, meadows, and farm fields. Camas Creek flows through for eight miles and supplies water to many of the lakes and ponds. Wheat, barley, and alfalfa raised here supplement natural feed for waterfowl. During March-April and October-November migrations, up to 100,000 ducks and 3,000 geese use the refuge on their way between breeding and wintering grounds. From June to August you will see large numbers of redhead, mallard, shoveler, lesser scaup, and teal ducklings and Canada geese goslings. The rare trumpeter swan also nests here. This is a great place to view raptors including northern harriers, red-tailed and Swainson's hawks, American kestrels, and great horned, long-eared, and short-eared owls. Observe bald eagles in winter and peregrine falcons, re-introduced on the refuge in 1983, occasionally in the summer. Look for colonies of herons, egrets, and ibis in the summer. Mudflats are good places to scan for shorebirds and large cottonwood groves at the entrance can act as "migrant traps" for songbirds. Frequently sighted mammals include muskrats, beavers, coyotes, jackrabbits, mule deer, and pronghorn. Moose are often seen in the willows along Camas Creek.

Directions: From Idaho Falls, drive about 36 miles north on Interstate 15 and turn east at Interchange 150 to Hamer. From Hamer, go north on a frontage road for about 2.5 miles. Turn west and go two miles (go over the Interstate 15 overpass) to the refuge headquarters. A network of roads on refuge dikes provides exceptional wildlife viewing from your vehicle. A brochure with a map and a bird checklist are available at the headquarters.

Ownership: USFWS (662-5423)
Size: 10,578 acres   **Closest Town:** Hamer

P   🚶

River otters are superbly adapted to their aquatic lifestyle. Their sleek bodies have thick tapering tails, small ears, short legs, and fully webbed feet. ALAN D. CAREY

84

76 | Market Lake Wildlife Management Area

Description: Extensive marshes and a canal are surrounded by farmland and elevated areas of sagebrush, grass, and basalt outcroppings on this WMA. Look for large numbers of waterfowl, especially snow geese and northern pintails during spring migration. Other common water birds include tundra swans, eared, western, and Clark's grebes, white-faced ibis, American avocets, black-necked stilts, and black-crowned night herons. In the emergent marshy vegetation look for marsh wrens, savannah sparrows, and yellow-headed blackbirds. Occasionally you can spot a bobolink, northern saw-whet owl, sora, American bittern, and Caspian tern. Elk and pronghorn winter on BLM land at the WMA's northern edge, and can sometimes be seen on the WMA. Cartier Slough, situated along the Henry's Fork, has similar species to Market Lake. A mixture of riparian vegetation, marshes, and upland areas provide good habitat for waterfowl, songbirds, and upland birds. Fishing and hunting are popular at both sites.

Directions: From Idaho Falls, follow Interstate 15 north for 16 miles to Roberts and turn east onto Interchange 135. At the signed "T" intersection, turn north and go 0.7 mile on a paved road. At this point follow a gravel road, which veers to the right and becomes paved again in a short distance. Follow this road to the end, viewing the Main and Triangle Marshes, which offer the best viewing. Follow the signs that lead to Sandy Marsh, the refuge headquarters, and East Spring. A map and bird checklist are available at the headquarters. To reach Cartier Slough, continue north on Interstate 15 to eight miles north of Roberts and exit eastward onto Idaho 33; Cartier Slough will be in 15 miles. Turn right to the parking lot just before the North Fork bridge.

Ownership: IDFG (522-7783), BLM
Size: 5,000 acres **Closest Town:** Roberts

Identified as distinct from the western grebe by the extent of its black crown, the Clark's grebe has just recently been recognized as a distinct species. JAN L. WASSINK

Idaho's largest population of nesting bald eagles is located in the South Fork Snake River Canyon, but these spectacular predators may be observed at several sites throughout the year. DENNIS HENRY.

77 | South Fork Snake River

Description: The South Fork stretches some 60 miles from Palisades Dam to its confluence with the Henry's Fork. There are several boating access sites in the corridor, or you can view from highway turnouts. In the first nine miles below Palisades Dam the river runs through a narrow channel. It then spreads out and flows around several island complexes. View the 60-foot Fall Creek waterfall just upstream from the Swan Valley Bridge. Below the bridge, the river continues around a series of islands for two miles and enters a scenic canyon where vertical walls reach hundreds of feet high and cottonwood and conifer trees grow on the islands and banks. The impressive canyon scenery continues downstream toward Heise, where cliffs give way to a level flood plain. From here to the Henry's Fork confluence (18 miles) farmland flanks the corridor, but the river is sheltered by broad, dense cottonwood forest and associated shrubs—excellent examples of natural riparian vegetation. In fact, the USFWS has classified these cottonwood galleries as Idaho's most important fish and wildlife habitat site. The state's largest population of nesting bald eagles is located in the canyon as well as significant populations of ducks, Canada geese, great gray owls, peregrine falcons, and other raptors. Bald eagles, trumpeter swans, golden eagles, ducks, and geese are present in high concentrations throughout the corridor in winter, when great gray and other owl species can be found in the cottonwoods. In all, over 80 species of birds can be seen. Also watch for beaver, river otter, raccoon, muskrat, mule and white-tailed deer, elk (in winter), and moose. Mountain goats are occasionally seen in the Bear Gulch area in winter. Never approach heron or eagle nesting or roosting areas on foot; maintain a distance of at least 100 yards while viewing from a vehicle or boat.

Directions: To access the upstream reaches, follow U.S. 26 approximately 45 miles east from Idaho Falls to Swan Valley (part of the Teton Scenic Route). From Swan Valley, continue 10 miles to the boat access at the base of Palisades Dam. There is also boat access at Irwin, at the bridge over the South Fork, just before Swan Valley, and at Conant Valley. By car, drive along the southwest bank of the river without crossing the bridge, and continue on Forest Road 58. Fall Creek waterfall and campground are on this side. Easy downstream access points are near Lorenzo, Menan, Ririe, and Heise. For a more remote and scenic drive, take Forest Road 206 east of Heise, which follows the river for 12 miles. At its terminus a hiking/pack trail continues along the river for five miles. Also visit Cress Creek Nature Trail just two miles downstream of Heise on River Road. This is a 0.75-mile walk through several habitat types. Brochures are available at the Idaho Falls BLM office. Many roads are closed or restricted in winter when boating is only recommended downstream of Heise. Check with the Palisades Ranger District for details. A highly recommended boater's guide to the river is available from the Idaho Falls offices of the BLM, IDFG, and USFS. Also pick up a Targhee National Forest map while at the USFS. The guide contains a map and shows all boat and river access points, water hazard areas, travel routes, and campgrounds.

Ownership: BLM (529-1020), USFS (523-1412), USBR (483-4085), The Nature Conservancy, PVT
Size: 45 miles of river   **Closest Town:** Swan Valley, Ririe, Menan

78 | Tex Creek Wildlife Management Area

Description: Highly diverse vegetation types make up this WMA, including bitterbrush and amelanchier shrub steppe, low sagebrush, aspen, tall sagebrush, Douglas-fir and willow riparian. Some 3,000 mule deer, 2,000 elk, and 50 moose are present mainly in the winter. Upland birds include sharp-tailed, sage, ruffed, and blue grouse and the introduced gray partridge and chukar. Common raptors include red-tailed hawks, American kestrels, northern harriers, golden eagles, and northern goshawks. Nesting songbirds include the yellow-rumped warbler, common redpoll, song and vesper sparrow, gray-crowned rosy finch, cliff swallow, and northern flicker. During the fall watch for migrating American goldfinches, black-capped chickadees, chipping and white-crowned sparrows, horned larks, Bohemian waxwings, and pine and evening grosbeaks. Mammals here include white and black-tailed jackrabbits, Nutall's cottontail, coyote, badger, and red fox. May to September viewing is best. The WMA is closed to vehicles December to mid-April.

Directions: From Idaho Falls, follow U.S. 26 about 20 miles east, to two miles east of the Ririe turnoff. Turn right onto Ririe Reservoir Road and go about 12 miles to the WMA headquarters. All roads are dirt and a map is highly recommended. They are available at the IDFG office in Idaho Falls. Camping is allowed near Ririe Dam.

Ownership: IDFG (522-7783), USBR, BLM
Size: 27,000 acres **Closest Town:** Ririe

79 | Palisades Reservoir

Description: This high altitude reservoir surrounded by steep forested mountains is part of the Greater Yellowstone Ecosystem. It supports one of the few nesting bald eagle populations that did not decline during the DDT pesticide era. Early summer through fall is the best time to visit. Look for bald eagles, osprey, western grebes, and great blue herons over the lake. Along the shoreline at dawn and dusk watch for moose, elk, mule deer, and black bears. The highest number of waterfowl and shorebirds are present during the fall, when leaf colors range from bright yellow to crimson red in sharp contrast with the deep greens of the coniferous trees.

Directions: From Idaho Falls, take U.S. 26 east for 55 miles. The highway follows the east side of the reservoir. There are several Forest Service roads and trails to explore; maps are available at the dam office.

Ownership: USFS (523-1412), USBR (483-4085)
Size: 17,000 acres **Closest Town:** Swan Valley

80 Grays Lake National Wildlife Refuge

Description: This "lake" is actually a large, shallow marsh with dense bulrush-cattail vegetation and very little open water. Caribou Mountain, at 9,803 feet, provides a picturesque backdrop. The refuge is home to the world's largest breeding concentration of greater sandhill cranes. It was also a reintroduction site for endangered whooping cranes, whose eggs were hatched under sandhill crane foster parents. May and June are the best months to see a variety of wildlife. The marsh is a major producer of Canada geese and many species of diving and dabbling ducks. Franklin's gulls nest in large colonies of up to 40,000 birds. Grebes, bitterns, and elusive rails nest in bulrush while wet meadows, shallow water, and mudflats harbor curlews, snipes, willets, and phalaropes. Common mammals include muskrats, badgers, ground squirrels, moose, and mule deer. In autumn, an abundance of sandhill cranes gather before migrating to New Mexico, Arizona, and Mexico. During late September to early October 3,000 cranes may be present. Only a few bird species are winter residents. Some rare bird sightings include lark buntings, bobolinks, and peregrine falcons. Hiking is allowed only in the northern half of the refuge and only from October 10 through March 31. Roads encircling the refuge may be impassable in winter. Cross-country skiing is great for winter access; the same restrictions apply as for hiking.

Directions: From Soda Springs, travel north and east on Idaho 34 for 33 miles. Turn north at the refuge sign onto a gravel road that circles the marsh and provides outstanding wildlife viewing. The refuge headquarters, about three miles down the road, has a modern visitor room with maps, checklists, and interpretive exhibits on refuge features—especially the whooping crane. Adjacent to the headquarters is a high overlook that provides an expansive view of the refuge. The observation platform is complete with educational displays and a spotting scope.

Ownership: USFWS (574-2755), BLM
Size: 18,300 acres **Closest Town:** Wayan

A greater sandhill crane tends her eggs at Grays Lake, home to the world's largest nesting concentration of these large birds. Grays Lake was also a reintroduction site for the endangered whooping crane. HARRY ENGELS

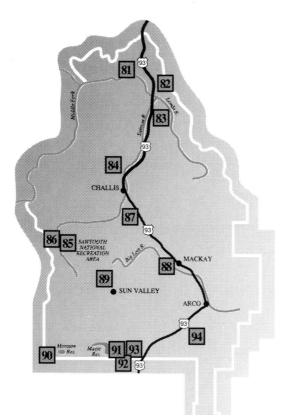

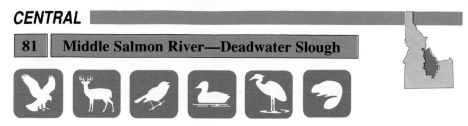

81 Middle Salmon River—Deadwater Slough

Description: The Deadwater Slough area is a complex of two large grassy meadows, braided stream channels, islands, and shallow backwater sloughs. Shrubby vegetation and cottonwoods line the river floodplain, with a surrounding habitat of steep canyon slopes covered with Douglas-fir and sagebrush. The great blue heron nests here in colonial rookeries while Canada geese nest on the islands and artificial structures and are often seen feeding in the large meadows and on several of the islands. In the upland areas, common birds include Lewis' and pileated woodpeckers, Nashville, yellow, and MacGillivray's warblers, yellow-breasted chat, Clark's nutcracker, rock wren, gray catbird, warbling vireo, song sparrow, and lazuli bunting. Look in the river for beaver and river otter. The Salmon River Canyon from North Fork downstream to Corn Creek (44 miles) is excellent for viewing wintering bald eagle, beaver, river otter, mule deer, elk, bighorn sheep, mountain goat, and very rarely mountain lion and bobcat. All but the mountain goat can be seen during the rest of the year, although in lower numbers. Corn Creek is the launch site for Middle Salmon River float trips (see site No. 26 in Region II). Beware of rattlesnakes during the summer and fall.

Directions: From North Fork on U.S. 93 head west on Salmon River Road (Forest Road 30). View along the river for the next 3.5 miles. In 0.6 mile from North Fork an inconspicuous dirt road turns off to the left to Newland Ranch Picnic Area. Park at the picnic area and cross the river on a rustic bridge closed to vehicle travel. An old road that eventually turns into a trail can be followed downstream two miles. Wildlife viewing is good all the way. Do not walk on the old road from February 15 to June 15 to avoid disturbing nesting geese. Do not walk on the islands from June 15 to August 1 to avoid disturbing the molting geese. During the summer and fall, canoeing, kayaking, rafting, or innertubing are all enjoyable ways to see the area. Put in at North Fork and take out at Deadwater Picnic Area. A ramp about 0.5 mile upstream from the Deadwater Picnic Area provides wheelchair access to the river. Winter access is by car. Maps and bird checklists are available from the USFS at North Fork and Salmon.

Ownership: USFS (756-2383)
Size: 3.5 miles of river **Closest Town:** North Fork

Even though riparian areas make up less than one percent of Idaho's land area, many animals depend on these areas for food, water, hiding cover, nesting sites, and migration routes.

82 | **Upper Salmon River—Salmon to North Fork**

Description: This 18-mile river stretch is a cottonwood riparian bottomland within a narrow valley flanked by the Beaverhead and Salmon River mountains. About halfway downriver is Tower Creek Bottoms, a privately-owned, 80-acre site that includes two ponds and a backwater slough with shrubby as well as open, park-like areas. View by canoe or from the main highway. Common spring and summer birds include the black-headed grosbeak, gray catbird, red-eyed vireo, Nashville and Wilson's warblers, yellow-breasted chat, ruffed grouse, wood duck, northern oriole, and pileated woodpecker. There is a great blue heron rookery nearby. Along the river look for osprey nest platforms that are used by osprey and Canada geese. Cliff swallows nest on river cliffs and bobolinks nest in adjacent pastures. Look for bald eagles in the winter and pronghorn, white-tailed, and mule deer, and river otter year-round.

Directions: From Salmon, head north on U.S. 93 and view for the next 18 miles. Rafters, kayakers, and canoers can use several small boat launches; inquire in Salmon at the BLM or USFS offices. Wagonhammer Springs, a small Forest Service campground with parking and picnic tables, is two miles before North Fork. A list of bird species is available from the BLM in Salmon. There are full facilities in Salmon and North Fork.

Ownership: PVT, BLM (756-5400), USFS (865-2383)
Size: 18 miles of river **Closest Town:** Salmon

Ruffed grouse males perch on logs and rocks to perform their "drumming" display. The rapid wing beat produces a distinctive sound like that of a chain saw or a tractor starting up. TIM CHRISTIE

Mountain lions occur statewide but are secretive and seldom seen. Years of pioneering research conducted through the University of Idaho has shed much light on the biology of these cats. TIM CHRISTIE

83 Lemhi River—Salmon to Leadore

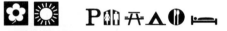

Description: This route travels through open sagebrush foothills and along large shrubby riparian zones of the Lemhi River. The route is interspersed with irrigated meadows and pastures. Over 100 bird species here are best viewed from May to July. Common species include great blue heron, Canada goose, great horned owl, American kestrel, red-tailed hawk, belted kingfisher, black-capped chickadee, warbling vireo, mountain bluebird, and song, vesper, Brewer's, and savannah sparrows. Less common are sandhill crane, short-eared owl, gray catbird, bobolink, lark bunting, Lewis' woodpecker, western wood-pewee, and common yellowthroat. Very few birds are present in fall and winter, but travelers at this time can look for rough-legged hawks, golden and bald eagles, and prairie falcons. Mule and white-tailed deer, pronghorn, and coyotes can be seen year-round.

Directions: From milepost 135 on Idaho 28, one mile south of Salmon, turn left onto an unnumbered county road that parallels the Lemhi River. The road is paved for a few miles then turns to a good gravel surface. Continue on this road for about 20 miles to Tendoy store, where it rejoins Idaho 28. Turn left onto the highway for 7.6 miles then turn left onto the county road again at Lemhi store. Stay on the road until Leadore, where it rejoins Idaho 28. There is a small BLM campground five miles south of Lemhi on Idaho 28. Salmon has full facilities. A map and bird checklist are available at the USFS office in Salmon and Leadore.

Ownership: BLM (756-5400), PVT
Size: 45 miles of river   **Closest Town:** Salmon

Yellow-bellied marmots occupy rocky areas where they feed on grasses and herbaceous plants. Basking in the sun is a common activity for marmots. JAN L. WASSINK

84 Morgan Creek

Description: Due to the diversity of habitats, Morgan Creek is an excellent place to view a variety of songbirds in the spring and summer. Common riparian birds include yellow and MacGillivray's warblers, warbling vireo, veery, lazuli bunting, willow flycatcher, song sparrow, and Brewer's and red-winged blackbirds. Major sagebrush species are the Brewer's and vesper sparrows, sage thrasher, and rock wren. In timbered areas the hermit thrush, yellow-rumped warbler, pine siskin, dark-eyed junco, ruby-crowned kinglet, and Clark's nutcracker are common. Uncommon birds here include Hammond's flycatcher, yellow-breasted chat, black-headed grosbeak, white-crowned and fox sparrows, gray catbird, northern oriole, and pileated woodpecker. Look for bighorn sheep, elk, pronghorn, and mule deer within the first five miles of the drive, among the sagebrush hills, from November to May.

Directions: From Challis, head north on U.S. 93 for 8.3 miles (to milepost 254.8) and turn left onto Forest Road 55 (gravel). View for the next 19.4 miles to Morgan Creek Summit. Five miles up the road is a BLM campground. Forest maps and a bird checklist are available from the USFS in Challis.

Ownership: BLM (756-5400), USFS (879-4321), PVT
Size: 19 miles of creek **Closest Town:** Challis

85 Sawtooth Fish Hatchery

Description: The Sawtooth hatchery features a room for viewing adult fish, an information center, and an observation platform. Visitors can see adult steelhead trout from March to May and the different life stages of chinook salmon year-round. The goal of the hatchery is to restore spring chinook salmon and steelhead trout to upper Salmon River tributaries. In the spring to early summer and in winter look for elk and deer in the surrounding fields. If you are visiting the area between March 15 and May 1 or late July to October 1, travel northeast of Stanley to Indian Riffles to view migrating and spawning steelhead (spring) and chinook salmon (fall) in a natural stream setting. Look also for common mergansers and great blue heron, which are often seen feeding on young fish.

Directions: The hatchery is six miles south of Stanley on Idaho 75 (one mile south of the Redfish Lake turnoff). It offers three guided tours daily from May to September. Indian Riffles is on Idaho 75, 13.5 miles east of Stanley.

Ownership: USFWS, managed by IDFG (774-3684)
Size: 100 acres **Closest Town:** Stanley

| 86 | Redfish Lake/Sawtooth Valley |

Description: This highly scenic valley is composed largely of sagebrush flats and grassy meadows bordered by lodgepole pine and aspen forests. Willows border the Salmon River and its tributaries, including Redfish Lake Creek. Sandhill cranes and savannah sparrows are often present in the meadows in the spring and summer. Willow nesters include white-crowned, Lincoln's, and song sparrows and Townsend's, Wilson's, and yellow warblers. Mule deer and coyotes are often seen in the valley while black bear and moose are rarely seen. Early summer wildflowers abound in the meadows. Redfish Lake is bordered by the spectacular, rugged Sawtooth Mountains on the upstream end and rimmed with coniferous forest on the remainder. The Lake has a visitor center with wildlife information and a boardwalk nature trail that traverses river otter and beaver habitat with active dams. A distant osprey nest is visible from the center. Birding is good on the lake and at the mouth of Redfish Lake Creek, where forest trails lead to high mountain lakes in the Sawtooth Wilderness. During spring and fall migration look on the lake for ring-billed gulls, numerous waterfowl species, and an occasional common loon. In timbered areas, ruby-crowned kinglet, hermit thrush, Cassin's finch, and yellow-rumped warblers are common throughout the summer. Hairy, pileated, black-backed, and three-toed woodpeckers are also found here. Frequently seen mammals are the yellow pine chipmunk, golden-mantled ground squirrel, red squirrel, and mule deer. In August and September, view kokanee salmon spawning in the creeks. Also observe bald eagles in the fall (and winter if there is open water) and the tracks of red fox and wolverine around the lake in winter.

Directions: From Stanley, go south on Idaho 75 for five miles to the Redfish Lake turnoff. Turn right and go two miles to the visitor center, open from mid-June to Labor Day, with maps, bird checklists, wildlife displays, interpretive programs, and hikes. The road to the lake is not plowed in the winter and is used for cross-country skiing and snowmobiling. Continuing south on Idaho 75, look for wildlife in the meadows and along the Salmon River while in the Sawtooth Valley.

Ownership: USFS (774-3681 or 726-7672)
Size: 1,600 acres **Closest Town:** Stanley

The Sawtooth hatchery produces three million chinook (king) salmon each year. A large chinook may grow to nearly five feet in length and weigh more than 100 pounds. It is the most highly prized of all Pacific salmon.

87 | East Fork Salmon River Canyon

Description: The roads paralleling the East Fork Salmon River and some of its feeder streams pass through both low and high elevational habitats. At 5,400 feet, the lower end of the river canyon is dominated by sagebrush and cottonwood riparian vegetation. The Boulder Creek road traverses Douglas-fir forest and willow/aspen riparian vegetation. By continuing on the Jim Creek road you will climb to alpine forests of Douglas-fir, subalpine fir, and whitebark pine bordering the White Cloud Peaks, a rocky mountain range with backcountry trail access to hundreds of lakes. The peaks are a proposed wilderness study area. At the end of the route you will be atop Railroad Ridge at 10,330 feet. In the lower, open areas look for chukars, red-tailed hawks, northern harriers, the occasional prairie falcon, pronghorn, and mule deer. Road Creek and Spar Canyon roads are good places to watch for the Challis wild horse herd. As you climb into alpine habitats look for elk, bighorn sheep, golden eagles, and black rosy finches. In most winters only the East Fork Road is open to travel, when visitors can see bighorn sheep, pronghorn, and mule deer.

Directions: *From Challis, go south on Idaho 75 for 17 miles and turn left onto East Fork Road. Viewing begins here. In about 3.5 miles you will pass Spar Canyon Road. In another 2.5 miles you will pass Road Creek Road. Continue 11 miles and turn right onto Big Boulder Creek Road (Forest Road 667). In about 4.5 miles you will reach the Jim Creek Road (Forest Road 669) where the road turns rough and narrow and is recommended for 4-wheel drive vehicles only. There is a trail at this junction into the White Cloud Peaks or you can continue up Jim Creek for five miles to the top of Railroad Ridge, the end of the route. In winter, only the East Fork Road is open for travel. Facilities listed are at the BLM East Fork Campground at the start of the route. Sawtooth and Challis National Forest maps are available in Stanley or Challis.*

Ownership: BLM (756-5400), USFS (774-3681 or 726-7672)
Size: 30-mile one-way drive **Closest Town:** Challis

The outstanding scenery of the Sawtooth National Recreation Area combined with nature trails and wildlife displays make Redfish Lake a popular spot for outdoor enthusiasts. GLENN OAKLEY

88 | Mackay Reservoir/Chilly Slough

Description: Mackay Reservoir is set in sagebrush grassland habitat. Exposed mudflats in the spring and fall attract numerous shorebirds and waterfowl, including cinnamon and green-winged teal, mallards, shovelers, pintail, scaup, and Canada geese. The very upper end of the shoreline still has mudflats when the reservoir is full. This land is private so view at this end by boat. Another close view is from the bluffs between U.S. 93 and the reservoir. To explore a more isolated area on foot, cross the dam and rocky cliffs to Black Daisy Canyon, where elk and mule deer winter. Chilly Slough is a palustrine emergent wetland in the Big Lost River Valley with the 12,662-foot Mt. Borah towering to the east. Visitors can walk along the edge of the marsh from the parking area for about 0.5 mile. From spring through fall the area is rich with birds, including the same water bird species as at Mackay along with willets, sandhill cranes, sora, marsh wrens, red-tailed hawks, golden eagles, and northern harriers. Tundra swans are sometimes present in the spring and fall. Several hundred pronghorn winter just east of U.S. 93 at Chilly Slough while in the summer they are more dispersed. You may also see mule deer from fall to spring between the reservoir and slough. Trout fishing is very popular at Mackay Reservoir and Chilly Slough supports spawning rainbow and brook trout. If you are heading to Challis between December and March, pull off the highway at Willow Creek Summit. Wintering elk can be viewed (use binoculars) feeding on curleaf mountain mahogany shrubs and bluebunch wheatgrass.

Directions: *To reach Mackay Reservoir, drive four miles north of Mackay on U.S. 93. To reach Chilly Slough, go 15 miles north of Mackay on U.S. 93. The parking turnout is 1.5 miles past Trail Creek Road. Parking for Willow Creek Summit is 10 miles north of Chilly Slough on the west side of U.S. 93. A bird checklist for the area is available from the BLM office in Salmon. Facilities listed are at Mackay Reservoir.*

Ownership: BLM (756-5400)
Size: 1,300 acres; 80 acres **Closest Town**: Mackay

Idaho's state bird, the mountain bluebird catches insects to feed it's young. Several hundred nest boxes placed along "bluebird trails" supplement natural cavities in trees.
TOM ULRICH

89 | Trail Creek/Corral Creek

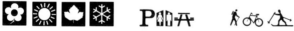

Description: These mountain creek corridors set in the upper reaches of a highly scenic, glacially carved valley are rich in wildlife due to their habitat diversity. Riparian corridors are lined with aspen, alder, mature cottonwood, and willows. Nearby forest stands are a mix of Douglas-fir and, at higher elevations, subalpine fir and limber and whitebark pine. Talus slopes, cliffs, sagebrush, and avalanche chutes add to the diversity. View from the road, where portions overlook riparian areas. Common wildlife include the red-tailed hawk, northern harrier, blue grouse, yellow-bellied sapsucker and other woodpeckers, American dipper, belted kingfisher, great blue heron, cinnamon teal and other ducks, red squirrel, Columbian ground squirrel, yellow-bellied marmot, beaver, elk, mule deer, and, rarely, moose. Look for active beaver dams and tell-tale rectangular holes made by pileated woodpeckers in large Douglas-fir and cottonwood trees. Although quite rare, mountain goats have been seen in the rock cliffs of upper Trail Creek. Elk are most visible in the winter when access is via cross-country skiing on Trail and Corral Creek Roads. There is a bluebird nest box trail along Corral Creek. Four miles in on Corral Creek, the Pioneer Cabin Trailhead has several loop trails from six to over 20 miles. Those traveling north of Ketchum can stop by the Sawtooth National Recreation Area's Headquarters and Visitor Center with excellent wildlife displays, slide shows, literature, and checklists. The Center is seven miles north of Ketchum on Idaho 75.

Directions: From Idaho 75 at Ketchum, turn northeast at the stoplight toward Sun Valley. Follow Trail Creek Road (Forest Road 408) 13 miles to Trail Creek Summit. In five miles you will pass the junction of Trail Creek Road and Corral Creek Road (Forest Road 137). Viewing is from the roads and short hikes off the roads. You can park at two overlooks along Trail Creek Road and at the Pioneer Cabin Trailhead. An area bird checklist is available at the USFS Ketchum Ranger District office, along with a list of species the FS would like reported if seen. Boundary Picnic Area with restrooms is about four miles past the Ketchum stoplight.

Ownership: USFS (622-5371)
Size: 17 miles of road   **Closest Town:** Sun Valley

 Idaho's lakes, marshes, and shorelines are home to fourteen species of colony nesting waterbirds.

90 | Camas Prairie Centennial Marsh

Description: In spring, most of this sprawling, 10,000-acre wet meadow is covered with shallow water, which starts to dry up in early July. Camas blooms in May cover the prairie with a blanket of purple. A long history of farming has drastically altered the prairie from its natural state. The marsh is a wetland aquisition and restoration project supported by Ducks Unlimited, The Nature Conservancy, Inc., and the IDFG. When completed, the marsh will encompass about 11,000 acres with an interpretive center and self-guided nature trails. On spring mornings the marsh is filled with thousands of waterfowl, shorebirds, and wading birds including snow and Canada geese, cinnamon, blue, and green-winged teal, canvasback, gadwall, Wilson's phalarope, willets, common snipe, killdeer, sandhill cranes, and long-billed curlews. Many of these species nest at the marsh. Eight raptor species frequent the area including the golden eagle, northern harrier, and red-tailed, rough-legged, and Swainson's hawks. The prairie is used as spring to fall range by mule deer and pronghorn. Elk winter on the prairie fringes. Several moose, introduced in 1986, appear to be thriving. Coyotes, red fox, badgers, striped skunks, and many small rodent species are found throughout the area. The best viewing is from spring to early summer. Hunting is popular in the fall. For additional viewing, Mormon Reservoir, five miles south of Fairfield, is home to a large ring-billed gull colony.

Directions: *There are east and west entrances to this site. From Fairfield on Idaho 20, go west 10 or 17 miles and turn south at the signs. You can drive, walk, or mountain bike the five miles of gravel roads. To reach Mormon Reservoir, turn south off Idaho 20, opposite the downtown Fairfield turnoff, and follow the signs.*

Ownership: IDFG
Size: 3,046 acres **Closest Town:** Fairfield

The Camas Prairie marshland provides important nesting habitat for several species of shorebirds, wading birds, raptors, and waterfowl, such as this mallard family. TIM CHRISTIE

91   Picabo Hills

Description: In this portion of Idaho's High Desert Range, look for large numbers of mule deer in early spring as they migrate north to their summer ranges. View sage grouse in the spring on their traditional display grounds, or leks, where the males will compete for status to attract females. In early spring, male grouse visit the site daily at dawn and stay for a few hours. The males will display for several weeks while the female will visit only for a short time near the end of March or early April to mate. Do not approach the grouse on foot; photographers should use telephoto lenses. Numerous lek sites are scattered along the route; scan open portions of the habitat during early mornings. An occasional elk, pronghorn, chukar, or gray partridge may also be viewed. Common sagebrush-dependent songbirds include the sage and Brewer's sparrow and sage thrasher. Red-tailed hawks, golden eagles, and other raptors are also common.

Directions: From Shoshone, go north on Idaho 75 for 20 miles and turn east at the Picabo Desert Road sign. This main desert road will take you on a 15-mile tour along the southern edge of Picabo Hills ending in the town of Picabo on U.S. 20. Numerous primitive roads can be taken off this main road, but their condition in early spring can be poor. A 4-wheel drive vehicle is recommended. Primitive roads can also be walked or mountain biked.

Ownership: BLM (886-2892), IDL, PVT
Size: 40,000 acres   **Closest Town:** Picabo, Shoshone

Standing with tail fanned and white neck feathers raised in a ruff, a male sage grouse displays on his breeding ground. He will inflate and compress air sacs in his throat to create loud popping sounds.
MICHAEL S. SAMPLE

92 Silver Creek Preserve

Description: This beautiful, meandering, spring-fed creek remains open all year and attracts an abundance of wildlife. Preserve land includes 15 miles of Silver Creek, its tributaries, and surrounding sagebrush desert. The Picabo Hills border the southern edge. View many waterfowl species and wading birds including sandhill crane, American bittern, and long-billed curlew. In sagebrush habitat look for sage grouse, vesper and Brewer's sparrows, and loggerhead shrikes. Several warblers can be seen in the willows and aspens along the creeks including orange-crowned, yellow, yellow-rumped, MacGillivray's, and Wilson's plus the yellow-breasted chat. Uncommon birds seen here during migration include the bald eagle, osprey, merlin, tundra swan, northern mockingbird, Caspian tern, and Virginia rail. Mammal inhabitants include striped skunk, river otter, beaver, mule deer, coyote, and porcupine. Elk visit in fall and winter. This site is world renowned for its trout fishing. Hiking, rafting, and canoeing are popular year-round while cross-country skiing is a favorite winter activity.

Directions: From Picabo on U.S. 20 drive west approximately four miles, just past the turnoff to Gannett. Turn left and go two miles, crossing a bridge, to the preserve headquarters, which will be on your right. Sign in and pick up a map of the preserve.

Ownership: The Nature Conservancy, Inc. (726-3007)
Size: 2,800 acres **Closest Town:** Picabo

93 Carey Lake Wildlife Management Area

Description: This shallow, marshy lake is surrounded by irrigated pasture, hay and grainfields, sagebrush, and lava rock outcrops. The primary viewing is from two parking areas (use binoculars or spotting scope) and by small boat. Tundra swan, Canada and snow goose, and many other waterfowl species are often seen here. Look for nesting Canada geese atop artificial platforms. Sandhill crane, American bittern, Virginia rail, American avocet, black-necked stilt, semi-palmated plover, willet, lesser and greater yellowleg, pied-billed grebe, and California gull are also regularly seen while white pelican, hooded merganser, black tern, and water pipit are uncommon. Fishing and waterfowl hunting are popular; only non-motorized boats are allowed on the lake.

Directions: From Carey on U.S. 20-26-93, travel east for one mile to a gravel road and parking area. The WMA is on the south side of the highway and visible from several turnoffs.

Ownership: IDFG (324-4350)
Size: 430 acres **Closest Town:** Carey

94 | Craters of the Moon National Monument

Description: This site's unusual "lunar" landscape is the result of the most recent lava flows on the Snake River Plain (only 2,000 years old). Sporadic volcanic eruptions occurring over the last 15,000 years have produced a mosaic of habitats, ranging from the barren lava flows of recent eruptions to the dense vegetation of older cinder cones covered by sagebrush and grass. The dramatic basaltic formations provide a unique setting for wildlife watching and photography. Observe mule deer, yellow-bellied marmots, pika, golden-mantled ground squirrels, red squirrels, blue and sage grouse, prairie falcons, and golden eagles in the spring and summer. Look for nesting violet-green swallows, ravens, and great horned owls near the openings of lava tube caves. The mountain bluebird, state bird of Idaho, commonly nests in cavities of limber pines that grow on the lava flows and cinder cones. Other common birds, present at various times of year, include the black-capped and mountain chickadees, hairy woodpecker, Clark's nutcracker, poor-will, and the yellow-bellied sapsucker. Reptiles warm themselves on the lava rock. Watch for racers, western rattlesnakes, gopher snakes, and sagebrush and short-horned lizards. June wildflower blooms can be spectacular. Fall visitors are likely to see mule deer and migratory songbirds. Though cross-country skiing is excellent here, winter months may not be very rewarding for viewing wildlife.

Directions: From Arco, take Interstate 93 southwest for 18 miles, where you will see the monument entrance signs. Drop in at the visitor center for maps and information about the wildlife, geology, and history of the area.

Ownership: NPS (527-3257)
Size: 53,546 acres   **Closest Town:** Arco

Reptile body temperatures fluctuate with environmental temperatures. This Mojave black-collared lizard basks on a sunny rock to warm itself. DON PAUL

SUPPORT IDAHO'S NONGAME WILDLIFE

Donations to the Idaho Nongame and Endangered
Wildlife Program may be made to
IDFG Nongame Trust Fund,
P.O. Box 25, Boise, ID 83707.

ABOUT DEFENDERS OF WILDLIFE

Defenders of Wildlife is a national nonprofit organization of
more than 80,000 members and supporters dedicated to preserving the natural abundance and diversity of wildlife and its
habitat. Defenders is working in close cooperation with
many state and federal agencies to develop state wildlife
viewing systems, including the state wildlife viewing guides.

If you are interested in becoming a member, annual dues are
$20, which includes six issues of our bimonthly magazine, *Defenders*. To join or
for further information, write or call Defenders of Wildlife, 1244 19th St. N. W.,
Washington, DC 20036, phone 202-659-9510

PLEASE HELP IMPROVE THE NEXT GUIDE

Since this guide is the first of its kind in Idaho, the sponsors welcome comments
concerning your experiences. Observations about the selection of sites, management
of the areas, adequacy of the facilities, need for interpretive information, usefulness of
the format, and site directions will all be considered when this guide is revised in the
future. Site evaluation forms are available upon request should you wish to recom-
mend changes or additional areas. Send letters to Defenders of Wildlife, 333 South
State Street, Suite 173, Lake Oswego, OR 97034. Your comments will be shared with
the sponsors.

MORE BOOKS FROM FALCON PRESS

Falcon Press publishes a wide variety of outdoor books and calendars, including the
state-by-state series of wildlife viewing guides called the Watchable Wildlife Series.
If you liked this book, please look for the companion books on other states.

If you want to know more about outdoor recreation in Idaho, then look for *The
Hiker's Guide to Idaho* and *Idaho—A State of Mind*—both from Falcon Press.

To purchase any of these books, please check with your local bookstore call toll-
free 1-800-582-BOOK. When you call, please ask for a free catalog listing all the fine
books and calendars from Falcon Press.

Falcon Press Publishing Co., Inc.,
P.O. Box 1718, Helena, MT 59624.